T0350404

QUANTITATIVE RISK ASSESSMENT

Quantitative risk assessments cannot eliminate risk, nor can they resolve trade-offs. They can, however, guide principled risk management and reduction – if the quality of assessment is high and decision-makers understand how to use it.

This book builds a unifying scientific framework for discussing and evaluating the quality of risk assessments and whether they are fit for purpose. Uncertainty is a central topic. In practice, uncertainties about inputs are rarely reflected in assessments, with the result that many safety measures are considered unjustified. Other topics include the meaning of a probability, the use of probability models, model uncertainty, how to understand and describe risk, the use of Bayesian ideas and techniques and the use of risk assessment in a practical decision-making context.

Written for professionals, as well as graduate students and researchers, the book assumes basic probability, statistics and risk assessment methods. Examples make concepts concrete, and three extended case studies show the scientific framework in action.

TERJE AVEN is Professor in Risk Analysis and Risk Management at the University of Stavanger, Norway, and a Principal Researcher at the International Research Institute of Stavanger (IRIS).

QUANTITATIVE RISK ASSESSMENT

The Scientific Platform

TERJE AVEN

University of Stavanger, Norway

CAMBRIDGE
UNIVERSITY PRESS

CAMBRIDGE
UNIVERSITY PRESS

University Printing House, Cambridge CB2 8BS, United Kingdom

One Liberty Plaza, 20th Floor, New York, NY 10006, USA

477 Williamstown Road, Port Melbourne, VIC 3207, Australia

314-321, 3rd Floor, Plot 3, Splendor Forum, Jasola District Centre, New Delhi - 110025, India

79 Anson Road, #06-04/06, Singapore 079906

Cambridge University Press is part of the University of Cambridge.

It furthers the University's mission by disseminating knowledge in the pursuit of education, learning and research at the highest international levels of excellence.

www.cambridge.org
Information on this title: www.cambridge.org/9780521760577

First published 2011

A catalogue record for this publication is available from the British Library

Library of Congress Cataloging in Publication data
Aven, T. (Terje)
Quantitative risk assessment : the scientific platform / Terje Aven.
p. cm.
Includes bibliographical references and index.
ISBN 978-0-521-76057-7 (Hardback)
1. Probabilities. 2. Risk assessment–Statistical methods.
3. Decision making–Statistical methods. I. Title.
QA273.A93 2011
519.2´87–dc22

2010037661

ISBN 978-0-521-76057-7 Hardback

Contents

Preface

Risk assessment is in many respects acknowledged as a scientific discipline per se: there are many master and PhD programmes worldwide covering this field, and many scientific journals and conferences highlighting the area. However, there are few books addressing the scientific basis of this discipline, which is unfortunate as the area of risk assessment is growing rapidly and there is an enormous drive and enthusiasm to implement risk assessment methods in organisations. Without a proper basis, risk assessment would fail as a scientific method or activity. Consider the following example, a statement from an experienced risk assessment team about uncertainty in quantitative risk assessments (Aven, 2008a):

The assessments are based on the "best estimates" obtained by using the company's standards for models and data. It is acknowledged that there are uncertainties associated with all elements in the assessment, from the hazard identification to the models and probability calculations. It is concluded that the precision of the assessment is limited, and that one must take this into consideration when comparing the results with the risk acceptance criteria and tolerability limits.

Based on such a statement, one may question what the scientific basis of the risk assessment is. Everything is uncertain, but is not risk assessment performed to assess the uncertainties? From the cited statement it looks like the risk assessment generates uncertainty. In any event, does this acknowledgment – that a considerable amount of uncertainty exists – affect the analyses and the conclusions? Only very rarely! My impression is that one writes such statements just to meet a requirement, and then they are put aside. This says a lot about the scientific quality of the assessments.

I strongly believe that the scientific platform of risk assessment – and quantitative risk assessment in particular – needs to be strengthened. The aim of this book is to contribute to this end. For many years I have been engaged in research trying to improve the scientific basis of risk assessment; I have written many books and papers related to the topic, and believe that the time has come to publish a fundamental exposition on the topic *quantitative risk assessment – the scientific platform*. The basic idea is to provide a framework for analysing and discussing the quality of the assessments using the scientific requirements of reliability and validity. The reliability requirement is concerned with the consistency of the "measuring instrument" (analysts, experts, methods, procedures), whereas validity is concerned with the assessment's success at "measuring" what one set out to "measure". This gives a new and original approach to the analysis and discussion.

The quality of risk assessment relates to the scientific building blocks of the assessments but also to the role of the assessments in the decision-making process. On an overall level one can say that the purpose of risk assessment is to support the decision-making – to adequately inform the decision-makers – but what type of decision support (knowledge, judgements) should the assessments provide? Are the objectives (expectations) accurate risk estimates and/or uncertainty characterisations (representations/expressions of the knowledge and lack of knowledge available)? The scientific quality of the assessments obviously needs to be seen in relation to these objectives. Also the requirements of reliability and validity depend on these objectives. Using these criteria, we can evaluate the quality of the assessments for different objectives of the assessments.

Uncertainty is a key topic when discussing the scientific platform of risk assessment. Other important issues are the meaning of a probability, the use of Bayesian ideas and concepts, the meaning of risk, how risk should be described, the meaning and use of models, model uncertainty, the meaning and use of probability models and parameters, and the value of information.

The book is general and is relevant for all types of applications, but safety engineering has the main focus.

For many years there has been a lively discussion about the scientific platform of statistical analysis in general: the Bayesian/non-Bayesian controversy; see e.g. Lindley (2000). However, there has not been much work on establishing a proper scientific basis for risk assessments. A number of papers address foundational issues of risk assessment; see e.g. Apostolakis (1988, 1990), Kaplan and Garrick (1981), Singpurwalla (1988, 2006) and Cooke (1991), but I am not aware of much work where fundamental scientific quality requirements such as reliability and validity are discussed in the context of a risk assessment (Aven and Heide, 2009).

Of the few contributions found in the literature I would like to draw attention to the first issue of the international scientific journal *Risk Analysis* in 1981, and in particular Weinberg (1981) and Cumming (1981). These authors describe some of the problems of risk assessments, and express a large degree of scepticism about the scientific reliability and validity of risk assessments. Weinberg notes that "one of the most powerful methods of science – experimental observations – is inapplicable to the estimation of overall risk". Graham (1995) writes that the discipline "should (and will) always entail an element of craft-like judgment that is not definable by the norms of verifiable scientific fact", and that "any determination that a risk has been 'verified' is itself a judgment that is made on the basis of standards of proof that are to some extent arbitrary, disputable, and subjective" (Aven and Heide, 2009).

I share many of the same views on the scientific basis of risk assessment. However, in order to gain more insight into this subject and be able to make guidance on how to ensure and strengthen the scientific quality of risk assessments, we need to clarify the scientific pillars of the risk assessment and tailor the assessments to the decision-making context. As mentioned above, this is exactly what the present book does.

Small illustrating examples are included in the book for making concepts concrete and to illustrate ideas and principles. Three extended examples (case studies) will be presented early in the book (Chapter 4) and are pursued through the rest of the book. The first of these examples is related to the analysis of accident data, the second relates to the siting of a Liquefied Natural Gas (LNG) plant and the third discusses the design of a safety system. The idea is not that every concept or step of a risk assessment would be illustrated in each case study, but that these cases would recur often enough that the readers get a feel for the overall scope and shape of a real risk assessment and its use and are able to relate the scientific requirements to these concepts and steps. The cases are simplified so that the intellectual lessons are clarified, but they are nevertheless realistic.

The three cases illustrate different types of risk assessments. The first case covers a statistical data analysis, whereas the second shows an example of a system analysis which is strongly based on modelling of the phenomena studied. In Case 2 a large number of unknown quantities (model parameters) on the subsystem/component level need to be assessed. The third case presents an example of a reliability analysis of a specific system. The results of such analyses constitute important input to risk assessments.

Before we present and analyse the three main cases, we first review basic concepts and perspectives on how to define, understand and describe risk (Section 2). The aim is to give the reader an overview of the many different ways one can look at risk and to provide a structure for the coming analysis. We also discuss some fundamental issues related to science in a risk assessment context (Chapter 3): what are the basic features of risk assessment as a scientific method and how is risk assessment related to other scientific disciplines? This chapter also summarises the reliability and validity requirements mentioned above. Then in Chapters 5 and 6 we examine the three cases with respect to these scientific requirements: Chapter 5 looks at the situation when the objective of the risk assessment is accurate risk estimation, whereas Chapter 6 restricts attention to situations where the objective of the risk assessment is uncertainty characterisations. In Chapter 7 we discuss the implications of the findings in Chapters 5 and 6 for risk management and communication. Key issues addressed are the use of risk acceptance criteria, risk reduction processes, and the cautionary and precautionary principles.

From this analysis we are led to Chapter 8 which discusses and provides guidance on how risk should be approached, i.e. how we should define, understand and describe risk, as well as use risk assessments in a decision-making context. Chapter 9 provides some conclusions from the previous chapters.

The book allows for scientific analysis of different types of risk assessments, in particular assessments which in a detailed way reflect human and organisational factors. The book includes examples of assessments which reflect such factors, but it is beyond the scope of the book to provide a detailed account of these types of assessments. Well-selected references are presented for readers who do want to delve deeper in this area. See Section 1.1.

The book is for professionals in the field, as well as for graduate students and researchers. It should also be of interest to many policy makers and business people. The book would make it possible for them to better understand the boundaries of risk assessments and how they should be used for decision-making. The book is advanced (conceptually) but at the same time rather simple and easy to read. It has been a goal to avoid too many technicalities, but without diminishing the requirement for precision and accuracy. The main ideas and principles are highlighted. Readers would benefit from a basic knowledge in probability calculus and statistics as well as in risk assessment methods. It has, however, been a goal to reduce the dependency on extensive prior knowledge. The key statistical and risk concepts will be introduced and discussed thoroughly in the book. Thus the readers do not need to be experts on, for example, regression analysis. The focus will be on the basic ideas – "advanced statistical analysis" is not required. Appendix A provides a summary of basic theory (e.g. probability, Bayesian analysis). Appendix B includes a listing of some key definitions.

Acknowledgments

Many people have provided helpful comments and suggestions for this book. In particular, I would like to acknowledge Eirik B. Abrahamsen and Roger Flage for the great deal of time and effort they spent on reading and preparing comments on earlier versions of the book. I would also acknowledge Jan Terje Kvaløy for his kind help in producing the probability distribution shown in Figure 6.2. I am also grateful to two anonymous reviewers for valuable comments and suggestions on the plans for the book.

This work has been funded by The Research Council of Norway through the SAMRISK and PETROMAKS research programmes. The financial support is gratefully acknowledged.

I also acknowledge the editing and production staff at Cambridge University Press for their careful and effective work.

1

Introduction to risk management and risk assessments. Challenges

This chapter provides a broad introduction to risk management and risk asssessment, as a basis for the analyses and dicussions in the coming chapters. The presentation highlights general features but also challenges related to the definitions and use of these tools. Key references for the chapters are Bedford and Cooke (2001), Vose (2008) and Aven and Vinnem (2007). The terminology is to a large extent in line with ISO (2009a). See summary of key definitions in Appendix B.

1.1 General features of risk management and risk assessments

Risk management is all coordinated activities to direct and control an organisation with regard to risk. Two main purposes of the risk management are to ensure that adequate measures are taken to protect people, the environment and assets from undesirable consequences of the activities being undertaken, and to balance different concerns, for example safety and costs. Risk management covers both measures to avoid the occurrence of hazards/threats and measures to reduce their potential consequences. In industries like nuclear and oil & gas, risk management was traditionally based on a prescriptive regulating regime, in which detailed requirements for the design and operation of the plant were specified (Kumamoto, 2007; Aven and Vinnem, 2007). This regime has gradually been replaced by more goal-oriented regimes, putting emphasis on what to achieve rather than on the means of doing so. Goal orientation and risk characterisations are two major components of these new regimes that have been enthusiastically endorsed by international organisations and various industries (see e.g. IAEA Guidelines (1995), HSE (2001), Kröger (2006); the IPCS and WHO risk terminology document (2004) and the risk management guidelines of the EU Commission (European Commission, 2000, 2003; IEC, 1993)). Such

1

an approach to risk management is believed to provide higher levels of performance both in terms of productivity and risk reduction (Aven and Renn, 2009b).

Quantitative risk assessment

Quantitative Risk Assessment (QRA) (also referred to as Probabilistic Risk Assessment − PRA) is a key tool used in these new approaches. A QRA systemises the present state of knowledge including the uncertainties about the phenomena, processes, activities and systems being analysed. It identifies possible hazards/threats (such as a gas leakage or a fire), analyses their causes and consequences, and describes risk. A QRA provides a basis for characterising the likely impacts of the activity studied, for evaluating whether risk is tolerable or acceptable and for choosing the most effective and efficient risk policy, for example with respect to risk-reducing measures. It allows for the calculation of expected values so that different risks can be directly compared. Common practice in probabilistic risk assessment avoids, however, the aggregation of the two components and leaves it to the risk evaluation or management team to draw the necessary conclusions from the juxtaposition of loss and probabilities (Aven, 2003; Kröger, 2005). In addition, second-order uncertainties are introduced via different types of uncertainty intervals to make the confidence of probability judgements more explicit (Apostolakis and Pickett, 1998; Aven, 2003), see also Sections 2.7 and 8.3. For some extensive reviews of the use of QRA/PRA in a historical perspective, see Rechard (1999, 2000).

Some of the basic tools used for analysing the probabilities and risk are statistical estimation theory, fault tree analysis (FTA) and event tree analysis (ETA). These tools belong to the following main categories of basic analysis methods:

(a) *Statistical methods:* Data are available to predict the future performance of the activity or system analysed. These methods can be based on data extrapolation or probabilistic modelling.
(b) *Systems analysis methods:* These methods (which include FTA and ETA) are used to analyse systems where there is a lack of data to accurately predict the future performance of the system. Insights are obtained by decomposing the system into subsystems/components for which more information is available. Overall probabilities and risk are a function of the system's architecture and of the probabilities on the subsystems/ component level (Paté-Cornell and Dillon, 2001).

Quantitative risk assessment (QRA) is often associated with system analysis methods (see e.g. Bedford and Cooke, 2001), but in this book we interpret QRA (PRA) as any risk assessment which is based on quantification of risk using probabilities.

A number of new and improved methods have been developed in recent years to better meet the needs of the analysis, in light of the increasing complexity of the systems and to respond to the introduction of new techno-logical systems (Aven and Zio, 2011). Many of the methods introduced allow for increased levels of detail and precision in the modelling of phenomena and processes within an integrated framework of analysis covering physical phenomena, human and organisational factors as well as software dynamics (e.g. Mohaghegh *et al.*, 2009; Luxhoj *et al.*, 2001; Ale *et al.*, 2009; Røed *et al.*, 2009). Other methods are devoted to the improved representation and analysis of the risk and related uncertainties, in view of the decision-making tasks that the outcomes of the analysis are intended to support. Examples of relatively newly introduced methods are Bayesian Belief Networks (BBNs), Binary Digit Diagrams (BDDs), multi-state reliability analysis, Petri Nets and advanced Monte Carlo simulation tools. For a summary and discussion of some of these models and techniques, see Bedford and Cooke (2001) and Zio (2009).

The traditional risk assessment approach used in QRAs can be viewed as a special case of system engineering (Haimes, 2004). This approach, which to a large extent is based on causal chains and event modelling, has been subject to strong criticism (e.g. Rasmussen, 1997; Hollnagel, 2004; Leveson, 2004). It is argued that some of the key methods used in risk assessments are not able to capture "systemic accidents". Hollnagel (2004), for example, argues that to model systemic accidents it is necessary to go beyond the causal chains – we must describe system performance as a whole, where the steps and stages on the way to an accident are seen as parts of a whole rather than as distinct events. It is not only interesting to model the events that lead to the occur-rence of an accident, which is done for example in event and fault trees, but also to capture the array of factors at different system levels that contribute to the occurrence of these events. Leveson (2007) makes her points very clear:

Traditional methods and tools for risk analysis and management have not been terribly successful in the new types of high-tech systems with distributed human and automated decision-making we are attempting to build today. The traditional approaches, mostly based on viewing causality in terms of chains of events with relatively simple cause-effect links, are based on assumptions that do not fit these new types of systems: These approaches to safety engineering were created in the world of primarily mechanical systems and then adapted for electro-mechanical systems, none of which begin to approach the level of complexity, non-linear dynamic interactions, and technological innovation in today's socio-technical systems. At the same time, today's complex engineered systems have become increasingly essential to our lives. In addition to traditional infrastructures (such as water, electrical, and ground transportation systems), there are increasingly complex communication systems, information systems, air transportation systems, new product/process development systems, production systems, distribution systems, and others.

Leveson (2004) argues for a paradigm-changing approach to safety engineering and risk management. She refers to a new alternative accident model, called STAMP (System-Theoretic Accident Modeling and Processes).

Nonetheless, the causal chains and event modelling approach has shown to work for a number of industries and settings. It is not difficult to point at limitations of this approach, but the suitability of a model always has to be judged with reference to not only its ability to represent the real world, but also its ability to simplify the world. All models are wrong, but they can still be useful to use a well-known phrase. Furthermore, the causal chains and event modelling approach is continuously improved, incorporating human, operational and organisational factors, as was mentioned above. Mohaghegh *et al.* (2009), for example, present a "hybrid" approach for analysing dynamic effects of organisational factors on risk for complex socio-technical systems. The approach links system dynamics, Bayesian belief networks, event sequence diagrams and fault trees.

For the purpose of the present book, it suffices to consider the basic analysis tools such as fault tree and event tree models, probability models and statistical inference based on these models.

Risk assessment covers risk analysis and risk evaluation; see Figure 1.1. Risk analysis is a methodology designed to determine the nature and extent of risk. It comprises the following three main steps:

1. Identification of hazards/threats/opportunities (sources)
2. Cause and consequence analysis, including analysis of vulnerabilities
3. Risk description, using probabilities and expected values.

This definition of risk analysis seems to be the most common, but there are others (refer to IRGC, 2005). One of these considers risk analysis as an overall concept, comprising risk assessment, risk perception, risk management, risk communication, and their interactions. This interpretation has been often used among members of the Society of Risk Analysis.

Expressing risk also means to perform sensitivity analyses. The purpose of these analyses is to show how sensitive the output risk indices are with respect to changes in basic input quantities, assumptions and suppositions.

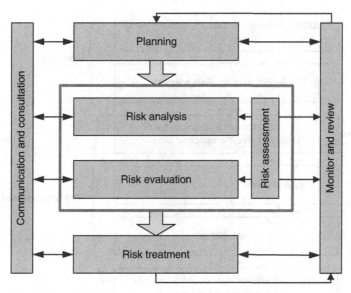

Figure 1.1   The risk assessment process (based on ISO, 2009b). Note that the ISO (2009a,b) does not include source identification as a part of risk analysis.

The sensitivity analyses can be used to identify critical systems, and thus provide a basis for selecting appropriate measures. To illustrate this, let R be a risk index, for example expressing the expected number of fatalities or the probability of a system failure, and let R_i be the risk index when subsystem i is in the functioning state. Then a common way of ranking the different subsystems is to compute the risk improvement potential (also referred to as the risk achievement worth) $I_i = R_i - R$, i.e. the maximum potential risk improvement that can be obtained by improving system i. The potential I_i is referred to as a risk importance measure. See Aven and Nøkland (2010) for a recent review of such measures.

Having established a risk description (risk picture), its significance is then evaluated (risk evaluation). Is the risk high compared to relevant reference values or decision criteria? How does alternative A compare with alternative B? etc. Risk analysis is often used in combination with risk acceptance criteria, as inputs to risk evaluation. Sometimes the term "risk tolerability limits" is used instead of risk acceptance criteria. The criteria state what is deemed as an unacceptable risk level. The need for risk-reducing measures is assessed with reference to these criteria. In some industries and countries, it is a requirement in regulations that such criteria should be defined in advance of performing the analyses.

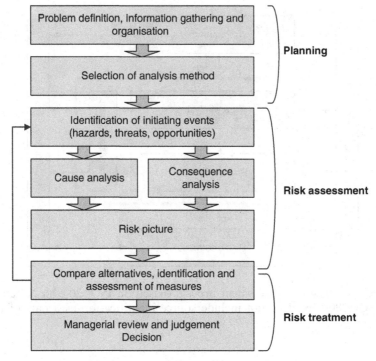

Figure 1.2   The main steps of the risk assessment process, covering the planning, the risk assessment and its use (based on Aven, 2008a).

The risk assessment process (planning, execution and use of risk assessments)

Risk assessment is followed by risk treatment, which is a process involving the development and implementation of measures to modify risk, including measures designed to avoid, reduce ("optimise"), transfer or retain risk. Risk transfer means sharing with another party the benefit or loss associated with the risk. It is typically effected through insurance.

"Planning" defines the basic frame conditions within which the risks must be managed and sets the scope for the rest of the risk assessment process. It means definition of suitable decision criteria as well as structures for how to carry out the risk assessment.

It is possible to detail the process in Figure 1.1 in many different ways to illustrate the planning, execution and use of risk analyses. Figure 1.2 shows an example based on Aven (2008a).

The results of the assessments need to be evaluated in the light of the premises, assumptions and limitations of these assessments. We refer to this stage of the process as the managerial review and judgement (Hertz and

Thomas, 1983; Aven, 2003). The assessments are based on some background knowledge that must be reviewed together with the results of the assessments. Consideration should be given to factors such as (Aven, 2003):

- which decision alternatives have been analysed
- which performance measures have been assessed
- the fact that the results of the analyses represent judgements (expert judgements)
- difficulties in assigning probabilities in the case of large uncertainties
- the fact that the assessments' results apply to models that are simplifications of the real world and real world phenomena.

The decision-making basis will seldom be in a format that provides all the answers that are important to the decision-maker. There will always be limitations in the information basis and the review and judgement described means that one views the basis in a larger context. Perhaps the analysis did not take into consideration what the various measures mean for the reputation of the enterprise, but this is obviously a condition that is of critical importance for the enterprise. The review and judgement must also cover this aspect.

The weight the decision-maker gives to the basis information provided depends on the confidence he/she has in those who developed this information. However, even if the decision-maker has maximum confidence in those doing this work, the decision still does not come about on its own. It is often difficult to make decisions when the risk is high. The decisions encompass difficult considerations and weighting with respect to uncertainties and values, and this cannot be delegated to those who create the basis information. It is the responsibility of the decision-maker to undertake such considerations and weighting, and to make a decision that balances the various concerns.

Apostolakis (2004, p. 518) makes this clear:

I wish to make one thing very clear: QRA results are *never* the sole basis for decision-making by responsible groups. In other words, safety-related decision-making is *risk-informed*, not risk-based.

Figure 1.3 illustrates the use of risk assessment in the decision-making. Risk assessment is carried out to support the decision-making, for example a choice between various concepts, design configurations, risk-reducing measures etc. Other types of assessment are also needed, such as cost-effectiveness analyses and cost–benefit analyses.

The same types of ideas are reflected in many other decision analysis frameworks and contexts, for example the analytic-deliberative process

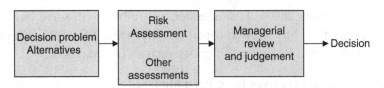

Figure 1.3   Model of the use of risk assessment to support decision-making.

recommended by the US National Research Council (1996) in environmental restoration decisions involving multiple stakeholders. According to this process, analysis "uses rigorous, replicable methods, evaluated under the agreed protocols of an expert community – such as those of disciplines in the natural, social, or decision sciences, as well as mathematics, logic, and law – to arrive at answers to factual questions"; while "deliberation is any formal or informal process for communication and collective consideration of issues. ... Participants in deliberation discuss, ponder, exchange observations and views, reflect upon information and judgements concerning matters of mutual interest and attempt to persuade each other." Such a process is particularly adapted to and relevant to decisions of great public interest.

Various decision-making strategies can form the basis for the decision. By "decision-making strategy" we mean the underlying thinking that goes on, and the principles that are to be followed with respect to how the decision is to be made, and how the process prior to the decision should be. Central to this is the question of who will be involved, how to use the various forms of analyses, and how the actual process is to be carried out.

ALARP principle

An example of such a strategy is to use risk acceptance (tolerability) criteria as inputs to risk evaluation. Another strategy is to adopt the ALARP principle, which means that risk should be reduced to a level that is as low as reasonably practicable. According to the ALARP principle, a risk-reducing measure should be implemented provided it cannot be demonstrated that the costs are grossly disproportionate relative to the gains obtained (the burden of proof is reversed). The standard approach when applying the ALARP principle, as for example used in the UK, is to consider three regions:

1. the risk is so low that it is considered negligible
2. the risk is so high that it is intolerable
3. an intermediate level where the ALARP principle applies.

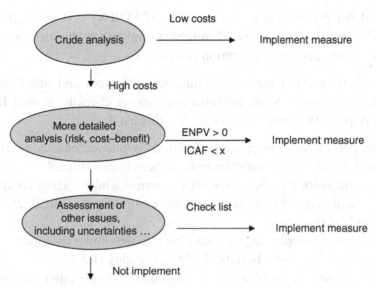

Figure 1.4   Procedure for implementing ALARP and the gross dispropor-tionate criterion (Aven and Vinnem, 2007).

In most cases in practice risk is found to be in region 3 and the ALARP principle is adopted. This will include a dedicated search for possible risk-reducing measures and a subsequent assessment of these in order to determine which to be implemented.

To verify ALARP, procedures mainly based on engineering judgements and codes are used, but also traditional cost–benefit analyses and cost-effectiveness analyses. When using such analyses, guidance values as above are often used to specify what values define "gross disproportion".

Conclusions are often self-evident when computing indices such as the expected cost per expected number of lives saved. For example, a strategy may be that measures will be implemented if the expected cost per expected number of lives saved (Implied Cost of Averting a Fatality – ICAF) is less than €2 million. Figure 1.4 sketches the main ideas of a procedure for how to implement ALARP and the gross disproportionate criterion in practice presented in Aven and Vinnem (2007).

The procedure can be summarised as follows:

- Perform a crude qualitative analysis of the benefits and burdens of the risk-reducing measure. If the costs are not judged to be large, implement the measure. Gross disproportion has not been demonstrated.
- If the costs are considered large, quantify the risk reduction and perform an economic analysis as indicated above (computing for example ICAF or the

expected net present value, i.e. E[NPV]). If E[NPV] > 0 or ICAF is low (typically less than some few € millions), implement the measure. Gross disproportion has not been demonstrated.

If these criteria are not met, assess uncertainty factors and other issues of relevance not covered by the previous analyses. A checklist is used for this purpose. Aspects that could be covered by this list are:

- Is there considerable uncertainty (related to phenomena, consequences, conditions) and will the measure reduce these uncertainties?
- Does the measure significantly increase manageability? High competence among the personnel can give increased assurance that satisfactory outcomes will be reached.
- Is the measure contributing to obtaining a more robust solution?
- Is the measure based on best available technology (BAT)?
- Are there unsolved problem areas: personnel safety-related and/or work environment-related?
- Are there possible areas where there is conflict between these two aspects?
- Is there a need for strategic considerations?

If the risk-reducing measure scores high on these factors (many yes answers), gross disproportion has not been demonstrated.

- Otherwise, the costs are in gross disproportion to the benefits gained, and the measures should not be implemented.

Cautionary and precautionary principles

The ALARP principle can be considered as a special case of the cautionary principle which states that in the face of uncertainty and risk, *caution* should be a ruling principle, for example by not starting an activity, or by implementing measures to reduce risks and uncertainties (HSE, 2001; Aven and Vinnem, 2007, p. 34). This principle is being implemented in all industries through safety regulations and requirements. For example, in the Norwegian petroleum industry it is a regulatory requirement that the living quarters on an installation should be protected by fireproof panels of a certain quality, for walls facing process and drilling areas. This is a standard adopted to obtain a minimum safety level. It is based on established practice of many years of operation of process plants. A fire may occur; it represents a hazard for the personnel and in the case of such an event, the personnel in the living quarters should be protected. The assigned probability for the living quarter on a specific installation being exposed to fire may be judged as low, but we know that fires occur from time to time in such plants. It does not matter whether we calculate a fire probability of x or y, as long as we consider the risks to be significant; and this type of risk has been judged to be significant by the authorities. The justification is experience from similar plants and sound judgements. A fire may occur, it is not an unlikely event, and we should then be prepared. We need no references to cost–benefit analysis. The requirement is based on cautionary thinking.

Risk analyses, cost–benefit analyses and similar types of analyses are tools providing insights into risks and the trade-offs involved. But they are just tools – with limitations. Their results are conditioned on a number of assumptions and suppositions. Being cautious also means reflecting this fact. We should not put more emphasis on the predictions and assessments of the analyses than that which can be justified by the methods being used.

The cautionary principle is implemented by for example:

- implementing robust design solutions, such that deviations from normal conditions are not leading to hazardous situations and accidents;
- designing for flexibility, meaning that it is possible to utilise a new situation and adapt to changes in the frame conditions;
- implementing safety barriers, to reduce the negative consequences of hazardous situations if they should occur, for example a fire;
- improving the performance of barriers by using redundancy, maintenance/ testing, etc.;
- applying quality control/quality assurance;
- adopting the precautionary principle, saying that if the consequences of an activity could be serious and subject to scientific uncertainties, then precautionary measures should be taken or the activity should not be carried out;
- implementing the ALARP principle.

The level of caution adopted will of course have to be balanced against other concerns such as costs. However, all industries would introduce some minimum requirements to protect people and the environment, and these requirements can be considered justified by reference to the cautionary principle.

We consider the precautionary principle a special case of the cautionary principle, as it is applicable in cases of *scientific uncertainties* about the possible consequences of the activity being considered. The distinction between the cautionary principle and the precautionary principle is adopted by the Health and Safety Executive in UK (HSE 2001) but is not so common in the literature. However, we find it useful for separating what are attitudes and actions in the case of risks and uncertainties, and what are attitudes and actions in the special case of scientific uncertainties. Many researchers and also lay people seem to use the term "precautionary principle" for both cases.

Defence-in-depth, robustness and resilience

In nuclear engineering and nuclear safety it is referred to a *defence-in-depth* philosophy meaning more or less the same as the cautionary principle. According to NRC (2010) the defence-in-depth is "an approach to designing and operating nuclear facilities that prevents and mitigates accidents that release radiation or hazardous materials. The key is creating multiple independent and redundant layers of defence to compensate for potential human and mechanical failures so that no single layer, no matter how robust, is exclusively relied upon. Defence-in-depth includes the use of access controls, physical barriers, redundant and diverse key safety functions, and emergency response measures."

It is prudent to distinguish between management strategies for handling the risk agent (such as a chemical or a technology) from those needed for the risk-absorbing system (such as a building, an organism or an ecosystem) (IRGC, 2005); see also Aven and Renn (2009b). With respect to risk-absorbing systems, robustness and resilience are two main categories of strategies/principles in the case of large uncertainties. Both strategies can be viewed as cautionary principles. *Robustness* (antonym: *vulnerability*) refers to the insensitivity of performance to deviations from normal conditions. Measures to improve robustness include inserting conservatisms or safety factors as an assurance against individual variation, introducing redundant and diverse safety devices to improve structures against multiple stress situations, reducing the susceptibility of the target organism (example: iodine tablets for radiation protection), establishing building codes and zoning laws to protect against natural hazards as well as improving the organisational capability to initiate, enforce, monitor and revise management actions (high reliability, learning organisations).

With respect to risk-absorbing systems, an important objective is to make these systems resilient so they can withstand or even tolerate surprises. In contrast to robustness, where potential threats are known in advance and the absorbing system needs to be prepared to face these threats, *resilience* is a protective strategy against unknown or highly uncertain events. Instruments for resilience include the strengthening of the immune system, diversification of the means for approaching identical or similar ends, reduction of the overall catastrophic potential even in the absence of a concrete threat, design of systems with flexible response options and the improvement of conditions for emergency management and system adaptation. Robustness and resilience are closely linked but they are not identical and require partially different types of actions and instruments. See also Chapter 2.

The decision-making strategy is dependent on the decision-making situation. The differences are large, from routine operations where codes and standards are used to a large extent, to situations with high risks, where there is a need for comprehensive risk-based information.

1.2 Challenges

In this book we are concerned with assessments and quantitative risk assessments in particular. Many researchers and analysts have questioned the scientific quality of these assessments (Aven, 2010a,h,i). For example, O'Brien (2000) argues that risk assessments generally serve the interests of business (i), as well as government agencies (ii) and many risk analysts (iii). She writes:

(i) The risk assessment gives the industry the aura of being scientific. The risk assessments show that the activities are safe, and most of us would agree that it is rational to base our decision-making on science. The complexity of a risk assessment makes it difficult to understand its premises and assumptions if you are not an expert in the field. In a risk assessment there is plenty of room for adjustments of the assumptions and methods to meet the risk acceptance criteria.

In the case of large uncertainties in the phenomena and processes studied, the industry takes advantage of the fact that in our society safety and environment-affecting activities and substances are considered innocent until "proven guilty". It takes several years to test for example whether a certain chemical causes cancer, and the uncertainties and choice of appropriate risk assessment premises and assumptions allow interminable haggling.

(ii) Risk assessment processes allow governments to hide behind "rationality" and "objectivity" as they permit and allow hazardous activities that may harm people and the environment (O'Brien, 2000, p. 106). The focus of the agencies is then more on whether a risk assessment has been carried out according to the rules, than on whether it provides meaningful decision support.

(iii) Risk analysts know that the assessments are often based on selective information, arbitrary assumptions and enormous uncertainties. Nonetheless they accept that the assessments are used to conclude on risk acceptability.

This critique of risk assessment is supported by many other researchers, see e.g. Reid (1992), Stirling (1998, 2007), Renn (1998), Tickner and Kriebel (2006) and Michaels (2008). Reid (1992) argues that the claims of objectivity in risk assessments are simplistic and unrealistic. Risk estimates are subjective, and there is a common tendency of underestimation of the uncertainties. The disguised subjectivity of risk assessments is potentially dangerous and open to abuse if it is not recognised. According to Stirling (2007), using risk assessment when strong knowledge about the probabilities and outcomes does not exist, is irrational, unscientific and potentially misleading. Renn (1998) summarises the critique drawn from the social sciences over many years and concludes that technical risk analyses represent a narrow framework that should not be the single criterion for risk identification, evaluation and management. Tickner and Kriebel (2006, pp. 53–55) and Michaels (2008) argue along the same lines as O'Brien (2000). Tickner and Kriebel (2006) particularly stress the tendency of decision-makers and agencies not to talk about uncertainties underlying the risk numbers. Acknowledging uncertainty can weaken the authority of the decision-maker and agency, by creating an image of being unknowledgeable. Precise numbers are used as a facade to cover up what are often political decisions. Michaels (2008) argues that mercenary scientists, including risk analysts, have increasingly shaped and skewed the technical literature, manufactured and magnified scientific uncertainty, and influenced government policy to the advantage of polluters and the manufacturers of dangerous products.

The answer to this critique is, according to O'Brien (2000), to look for an alternative to risk assessments. But in our view there is no alternative to risk assessments: to support the decision-making we need to assess risk. The right way forward is not to reject risk assessment, but to improve the tool and its use. This seems also to be the conclusion made by most researchers in the field. The challenge is how decision-making on risk can be informed by the best available technical and scientific knowledge (e.g. Stirling, 1998, p. 100; Apostolakis, 2004). We need to strengthen the quality of the risk assessments and the associated risk assessment process, to meet the above critique. However, to be able to do this we need to be precise on the fundamentals of the risk assessments and we need to establish a suitable framework for being able to make judgements about the scientific quality of the risk assessments. The aim of this book is to contribute to this end. By addressing the basic building blocks of the risk assessments, for example related to how to understand and describe risk and uncertainties, we are able to study the two fundamental scientific requirements: reliability and validity of the risk assessments. The reliability requirement is concerned with the consistency of the "measuring instrument" (analysts, methods, procedures), whereas validity is concerned with the assessment's success at "measuring" what one set out to "measure". This analysis also provides insights on how to manage risk and in particular how to define and use managerial review and judgement in a practical decision-making context.

We may all acknowledge that safety-related decision-making should be *risk-informed*, but practice shows that it is common to apply *risk-based* approaches. This may be a result of a more or less conscious management strategy but as we will see from the analysis in the coming chapters, it is strongly influenced by the adopted scientific approach to risk.

2

Concepts and perspectives on risk

In this chapter we first review and discuss common definitions of risk (Section 2.1–2.6). From this review and discussion we define the concepts and perspectives of probability and risk that will be used as the basis for the risk assessments studied in Chapters 5–7 (Sections 2.7–2.8). These perspectives specify not only how risk is defined, but also how to express risk. We also include an illustrative example (Section 2.9). Basically we will distinguish between two main lines of thinking, one where risk is considered an "object-ive" property of the activity studied and the aim of the risk assessment is to accurately estimate this risk, and one where uncertainty is a main component of risk and the aim of the risk assessment is to describe the uncertainties. The final Section 2.10 provides a summary of key concepts and perspectives. Some basic references for this chapter are Aven (2009a,b, 2010a,e).

2.1 Risk equals expected value

The concept of risk is defined in many ways. In engineering contexts, risk is often linked to the expected loss; see e.g. Lirer *et al.* (2001), Mandel (2007), Verma and Verter (2007) and Willis (2007). However, such an understanding of risk means that there is no distinction made between situations involving potential large consequences and associated small probabilities, and frequently occurring events with rather small consequences, as long as the sums of the products of the possible outcomes and the associated probabilities are equal. For risk management these two types of situations normally would require different approaches. In general expected value decision-making is misleading for rare and extreme events (Haimes, 2004; Aven, 2010a). The expected value does not adequately capture events with low probabilities and high conse-quences. Take as examples nuclear accidents and terrorism risk, where the possible consequences could be extreme and the probabilities are relatively low.

The expected value can be small, say 0.01 fatalities, but extreme events with millions of fatalities may occur, and this needs special attention.

Granger Morgan and Henrion (1990) provide a detailed account of the need for seeing beyond expected values in risk management. They point to common decision analysis frameworks, and in particular the expected utility theory (see Section 2.4). These frameworks reflect the need for seeing beyond expected values in risk management by incorporating the decision-maker's *risk aversion* or *risk-seeking attitude* in the utility (loss) function used. Risk aversion means that the decision-maker's certainty equivalent is less than the expected value; the certainty equivalent is the amount of payoff (e.g. money or utility) that the decision maker has to receive to be indifferent between that payoff and the actual "gamble". A risk-seeking attitude means that the decision-maker's certainty equivalent is higher than the expected value (Levy and Sarnat, 1994). Only in the case that the decision-maker is risk neutral, can expected values replace the information provided by the whole probability distributions. This does not mean however that the decision-maker in such a case would consider the risk management response only based on the subjectively assigned expected values. Cautionary measures may be considered justified also in this case to meet aspects (surprises) not covered by the expected utility analysis.

This analysis points to the difference between believing that a probability distribution can be safely summarised via its expected value and believing that the expected value is all that matters. The former conception is linked to the decision-maker's risk neutrality, whereas the latter conception often is a result of traditional statistical thinking where the law of large numbers provide support for using expected values, and uncertainties beyond these values are considered to be randomness of less importance for the decision-making (Aven, 2010i).

We conclude that we need to see beyond expected values when addressing risk. This is also reflected in the ways risk is most commonly defined in standards and in the scientific literature as discussed in the next section.

2.2 Risk is defined through probabilities

Some of the most typical definitions of risk in engineering contexts are:

1. Risk is a measure of the probability and severity of adverse effects (Lowrance, 1976).
2. Risk is the combination of probability and extent of consequences (Ale, 2002).
3. Risk is equal to the triplet (s_i, p_i, c_i), where s_i is the ith scenario, p_i is the probability of that scenario, and c_i is the consequence of the ith scenario, $i = 1, 2, \ldots N$ (Kaplan and Garrick, 1981; Kaplan, 1991).

What is common for all these definitions is that the concept of risk comprises events (initiating events, scenarios), consequences (outcomes) and probabilities. Uncertainties are expressed through probabilities. Severity is a way of characterising the consequences. It refers to intensity, size, extension, scope and other potential measures of magnitude, and affects something that humans value (lives, the environment, money, etc.). Losses and gains, for example expressed by money or the number of fatalities, are ways of defining the severity of the consequences. We formalise these definitions by writing

$$\text{Risk} = (A, C, P),$$

where A represents the events (initiating events, scenarios), C the consequences of A, and P the associated probabilities. Examples of events A are: gas leakage occurring in a process plant, and the occurrence of a terrorist attack. Examples of C are the number of casualties due to leakages, terrorist attacks, etc.

This definition of risk is, however, not meaningful without an interpretation of the probability P. Basically there are two ways of interpreting a probability (Bedford and Cooke, 2001; Aven, 2003; Appendix A):

(a) A probability is interpreted as a relative frequency P_f: the relative fraction of times the event occurs if the situation analysed were hypothetically "repeated" an infinite number of times; P_f is referred to as a frequentist probability. It can be understood as a parameter of a probability model, see Appendix A.

(b) Probability P is a subjective measure of uncertainty about future events and consequences, seen through the eyes of the assessor and based on some background information and knowledge (the Bayesian perspective). The probability is referred to as a subjective or knowledge-based probability.

A probability can also be given other interpretations (Singpurwalla, 2006; Section 2.3), but for practical use in a risk context we see no alternatives to (a) and (b).

Following definition (a) we produce estimates of the underlying "true" risk. This estimate is uncertain, as there could be large differences between the estimates and the correct risk values.

Kaplan and Garrick (1981) (see also Abramson (1981) and Apostolakis (1990)) meet this challenge by introducing subjective probabilities P, expressing the analysts' (experts') epistemic uncertainty about the relative frequencies P_f. The approach is referred to as the probability of frequency approach.

The variation in the outcomes of the "experiment" that for example generates the true value of P_f, is often referred to as aleatory (stochastic) uncertainty.

Following the Bayesian approach (b), we assign a probability by performing uncertainty assessments, and there is no reference to a correct probability. A probability is always conditional on a background knowledge, and given this background knowledge there are no uncertainties related to the assigned probability, as it is an expression of uncertainty.

If we assign a probability of 0.4 (say) for an event A, we compare our uncertainty (degree of belief) of A occurring with a standard event, for example drawing a red ball from an urn having 10 balls where 4 are red. The uncertainty (degree of belief) about A and the standard event is the same. The assignments are judgements based on the assessors' background knowledge, which we denote by K. To show the dependency on K, we write $P(A \mid K)$, where A is the event of interest. The background knowledge could be based on hard data and/or expert judgements. Assumptions are also included, for example related to the use of specific models. The background knowledge needs to be reported along with the assigned probabilities.

A betting interpretation of subjective probabilities also exists (Singpurwalla, 2006), but is not used in the present analysis. We refer to the discussion in Section 8.2 for some comments concerning this betting interpretation.

The term "subjective probability" is often problematic to use in real-life applications, as the word "subjective" is considered non-scientific and arbitrary, as noted by for example Lindley (2000) and North (2010). Another term is required. This need is also motivated by the fact that the term "subjective probability" is commonly linked to the betting interpretation which is problematic to use in a risk and uncertainty analysis context (refer to Section 8.2). Instead of subjective probabilities it is common to use terms such as "judgmental probability" (North 2010) and "knowledge-based probability" (Aven, 2010b). In this book we use both the terms "subjective probability" and "knowledge-based probability".

Note that the definitions 1–3 above can be applied for both interpretations of a probability, i.e. (a) and (b).

Probabilities are used as a tool to express uncertainties, but how good is this tool? Consider the probability P(attack), where the event "attack" is related to a specific period of time and location. Say that the assigned probability equals 0.01. Does this number provide an informative description of the uncertainties related to the event "attack"? No, would be the clear answer from many risk researchers and analysts. Uncertainties beyond the probabilities should be taken into account. Several definitions of risk have been suggested which are in line with this thinking, as shown in the following section.

2.3 Risk is defined through uncertainties

Examples of definitions of risk where uncertainty is an explicit component or an essential feature of the concept of risk are:

4. Risk refers to uncertainty of outcome, of actions and events (Cabinet Office, 2002).
5. Risk is a situation or event where something of human value (including humans themselves) is at stake and where the outcome is uncertain (Rosa, 1998, 2003).
6. Risk is an uncertain consequence of an event or an activity with respect to something that humans value (IRGC, 2005).
7. Risk is equal to the two-dimensional combination of events/consequences and associated uncertainties (Aven, 2007a, 2010e).
8. Risk is uncertainty about and severity of the consequences (or outcomes) of an activity with respect to something that humans value (Aven and Renn, 2009a).

A quick look at 4–8 may give the impression that they are not that different from 1–3. However, there are important principal differences, as will be clear from the coming analysis. Probability is just a tool used to represent or express the uncertainties. The thesis of all the perspectives and definitions 4–8 is that risk should not be limited to (A,C,P). The uncertainties should be highlighted. But what does this mean? Below (Sections 2.5–2.9) we will discuss this issue and provide a formal structure (framework) for the perspectives and definitions 1–8. The structure distinguishes between the risk concept on the one hand and risk descriptions on the other, and clarifies the meaning of fundamental concepts such as second-order probabilities and uncertainties. Analogous to the (A,C,P) perspective, we define an (A,C,U) risk perspective where uncertainty U replaces probability P. The perspectives 7 and 8 are expressing more or less the same and are in line with this (A,C,U) definition. How definitions 4–6 relate to the (A,C,U) perspective will be explained in Section 2.5.

2.4 Other definitions of risk

There are also other ways of understanding risk. We would like to mention two definitions which are often referred to in the economic risk and decision analysis field. The first defines risk by the expected disutility, i.e – Eu(C), where C is the outcomes (consequences) and u(C) the utility function (Campbell, 2005). Among classical decision theorists the expected (dis)utility provides the basis for rational choices. According to this definition, the preferences of the decision-maker are a part of the risk concept. The result is a mixture of scientific assessments of uncertainties about C and the decision-maker's preferences and values concerning different values of C and the associated probabilities. In our view, and this view is shared by many risk experts, the preferences and values should not be a part of the risk concept and the risk assessments (Paté-Cornell, 1996). There will be a strong degree of arbitrariness in the choice of the utility function, and some decision-makers would also be reluctant to specify the utility function as it reduces their flexibility to weight different concerns in specific cases. Risk should be possible to describe also in case the decision-maker is not able or willing to define his/her utility function.

In the second definition, risk refers to situations with objective probabilities for the randomness the decision-maker is faced with (Douglas, 1983). In economic applications a distinction has traditionally been made between risk and uncertainty: under risk the probability distribution of the performance measures can be assigned objectively, whereas under uncertainty these probabilities must be assigned or estimated on a subjective basis. This perspective goes back to Knight (1921). The risk concept then expresses variation in populations and is referred to as aleatory uncertainty; cf. Section 2.2 (Paté-Cornell, 1996). Although this definition is often referred to, it is not so often used in practice. The problem is of course that we seldom have known, objective distributions, and then we cannot refer to the risk concept. The Knightian definition violates the intuitive interpretation of risk (Vercelli, 1995; Holton, 2004), which is related to situations of uncertainty and lack of predictability, and is in general inconsistent with all the definitions 1–8. We will not look closer at these two definitions in this book.

2.5 Comparison of some common risk definitions and the (A,C,U) perspective

In this section we examine how the definitions 4–8 relate to the risk definition (A,C,U) and its corresponding risk description. According to 4, risk refers to uncertainty of outcome, of actions and events (Cabinet Office, 2002). Hence strictly speaking risk is not (A,C,U) but only U.

As an example, consider the number of fatalities in traffic next year in a specific country. Then the uncertainty is rather small, as the number of fatalities shows rather small variations from year to year. Hence, according to this definition of risk, we must conclude that the risk is small, even though the number of fatalities is many thousands each year. Clearly, this definition of risk fails to capture an essential aspect, the consequence dimension.

Uncertainty cannot be isolated from the intensity, size, extension etc. of the consequences. Take an extreme case where only two outcomes are possible, 0 and 1, corresponding to 0 and 1 fatalities, and the decision alternatives are r and s, having uncertainty (probability) distributions (0.5, 0.5), and (0.0001, 0.9999), respectively. Hence for alternative r there is a higher degree of uncertainty than for alternative s, meaning that risk according to definition 4 is higher for alternative r than for s. However, considering both dimensions, both uncertainty and the consequences, we would of course judge alternatives to have the highest risk as the negative outcome 1 is nearly certain to occur.

According to definition 5, risk is a situation or event where something of human value (including humans themselves) is at stake and where the outcome is uncertain (Rosa, 1998, 2003). Hence strictly speaking risk is A, and not as in our definition (A,C,U). However, Rosa expresses risk using the description (A,C,U), and refers to probability as a tool to describe the uncertainties. The Rosa (1998, 2003) definition is thoroughly discussed by Aven and Renn (2009a). The conclusion is that compared to common terminology, the Rosa definition leads to conceptual difficulties that are incompatible with the everyday use of risk in most applications. By considering risk as an event (A), we cannot conclude, for example, about the risk being high or low, or compare different options with respect to risk. The same conclusion is made for definition 6, which says that risk is an uncertain consequence of an event or an activity with respect to something that humans value (IRGC, 2005). This definition is similar to Rosa (1998, 2003)'s definition but the event A is replaced by the consequence C.

Definitions 7 and 8 are consistent with the (A,C,U) definition, although 8 introduces the term "severity" which refers to intensity, size, extension, scope and other potential measures of magnitude, and affects something that humans value (lives, the environment, money, etc.), as mentioned in Section 2.2. It is important to note that the uncertainties relate to the consequences (which include the events) – the severity is just a way of characterising the consequences.

2.6 The ontological status of the various risk concepts

The ontological status of the various definitions can be summarised as follows (Aven *et al.*, 2010). Risk defined by 4–8 exists "objectively", in the sense of "broad inter-subjectivity", as explained by the following arguments: The meaning of A or C we normally agree on, and no one (with normal senses) would dispute that future events and consequences are unknown.

"Being unknown" is not dependent on your knowledge about these events, it simply reflects that the future cannot be accurately foreseen.

A knowledge-based (subjective) probability is by definition subjective and dependent on the assigner. Inter-subjectivity could be obtained in some cases when the database is strong. But one cannot claim that risk definitions that are based on subjective probabilities are objective or broadly inter-subjective.

Modelling is also subjective, but in many cases natural model choices exist and inter-subjectivity is achieved. However, the established consensus model could be challenged by new knowledge. What was considered a truth is rejected in the light of new insights and evidence. Consequently the risk definitions which are based on relative frequencies are classified at best as inter-subjective.

2.7 A risk assessment perspective based on the (A,C,P_f) definition

This section introduces one of the two risk perspectives that we will look closer at in the coming chapters, the (A,C,P_f)-based perspective.

We consider an activity and make the following definitions of risk associated with this activity:

Risk $= (A,C,P_f)$, where P_f is a relative frequency-interpreted probability (or a related probability model parameter such as the expected number of occurrences of the event A per unit of time, where expectation is with respect to a relative frequency-interpreted probability).

As an example, consider the operation of a plant and the risk associated with accidents that could lead to fatalities. Then we can define events (such as gas leakages), associated consequences (losses), and probabilities, expressing for example the probability P_f of at least 10 fatalities next year. The P_f is interpreted as a property of the plant or, more precisely, the infinitely large population of similar plants that this particular plant belongs to. The frequentist probability P_f equals the fraction of plants in this population where an event occurs resulting in at least 10 fatalities.

Definitions 1–3 are covered by this risk perspective if the probabilities are relative frequency-interpreted.

In this case the risk is unknown as P_f is unknown. Risk assessment is introduced to describe (estimate) the risk. The description covers an estimate P_f^* of P_f, as well as assessments of uncertainties about P_f^* and P_f. Thus, if this perspective to risk is the starting point, we are led to a risk description:

Risk description in the (A, C, P_f) case $= (A, C, P_f^*, U(P_f^*), K)$, where $U(P_f^*)$ refers to an uncertainty description of P_f^* relative to the true value P_f, and K is the background knowledge that the estimate and uncertainty description is based on. We refer to $U(P_f^*)$ as a second-order uncertainty description. If we use subjective probabilities P to express our uncertainties about P_f, in line with the probability of frequency approach, the risk description takes the form:

Risk description according to the probability of frequency
approach $= (A,C,P_f^*,P(P_f),K)$,

where K now is the background knowledge that the estimate P_f^* and the probability distribution P is based on. Kaplan and Garrick (1981) refer to this description as the second-level definition of risk – it is combined with the first-level (A,C,P) definition (3) (see Section 2.2).

The probability P is a second-order probability. Möller *et al.* (2006) claim that the attitude of philosophers and statisticians towards second-order probabilities has been mostly negative, due to fears of an infinite regress of higher and higher orders of probability. This is hard to understand, as noted also by Sahlin (1993, p. 26). As stressed in Section 2.2, it has no meaning to talk about uncertainties of a subjective probability.

This issue has been strongly debated in the literature; see e.g. Gärdenfors and Sahlin (1988) and Sahlin (1993). According to Sahlin (1993), no one would seriously dispute that we have beliefs about our beliefs. We question this assertion. As a professional risk assessor, one is trained in the process of transforming uncertainty into probabilities. If the assessor assigns a probability of an event B equal to 0.3 based on a specific background knowledge, there is no reason why he/she should dispute his/her own assignment as it expresses his/her uncertainties (degree of belief). He/she may experience a precision problem, in particular when assessing events on the lower part of the probability scale. It could for example be difficult to distinguish between numbers such as 10^{-5} and 10^{-6}. However, the second-order probability issue is in our view not so much about having beliefs about beliefs, but the limitations of the probabilities in capturing the relevant uncertainty aspects. The second-order probabilities, i.e. the subjective probabilities, are based on some background knowledge and this knowledge could be wrong or poor in many respects. How should we reflect this in our risk description? Should we add an uncertainty component U in the risk description $(A,C,P_f^*,P(P_f),K)$, so that it becomes $(A,C,P_f^*,P(P_f),U,K)$? Yes, is the answer if we adopt the (A,C,U) perspective (as will be discussed in the next section). Then we need to see beyond the subjective probabilities P.

Although not mentioned so far, sensitivities constitute an integral part of the risk description. The sensitivities show how the output risk indices, for example P_f^* and $P(P_f)$, are influenced by changes in the background knowledge, in particular assumptions and suppositions.

The starting point for this analysis of the second-order probabilities was the risk concept (A,C,P_f) and following the probability of frequency approach we are led to the Bayesian framework. In cases of observations X, Bayesian updating of the subjective probabilities P are carried out using the standard Bayesian machinery going from a prior distribution to the posterior distribution. Bayesian theorists would not, however, refer to the P_f values as probabilities, but chances or propensities. The Bayesian framework is introduced in Appendix A.2.2 and explained in more detail in Chapter 6.

A pure traditional statistical approach would not allow for subjective probabilities P. The uncertainty $U(P_f^*)$ in the risk description would only reflect statistical variation, and could be expressed using a confidence interval. The degree of relevancy of the data would then not be taken into account. For this approach we obtain a risk description:

Risk description according to the pure traditional statistical
approach $= (A,C,P_f^*,d(P_f),K)$,

where d is a traditional confidence interval for P_f.

Vulnerability and resilience

Consistent with this risk perspective we may define the concept vulnerability (Aven, 2008a):

Vulnerability (antonym robustness) $= (C, P_f \mid A)$,

in other words, the vulnerability is the two-dimensional combination of consequences C and associated relative frequency-interpreted probabilities, given the occurrence of an initiating event A. For example, the vulnerability of a person with respect to a specific virus is the potential consequences of this virus and associated frequentist probabilities. The vulnerability description follows the same logic as that of risk:

Vulnerability description in the $(C,P_f \mid A)$ case $= (C,P_f^*,U(P_f^*),K \mid A)$.

When we say that a system is vulnerable, we mean that the vulnerability is considered high. The point is that we assess the combination of consequences and probability to be high should the initiating event A occur. If we know that the person is already in a weakened state of health prior to the virus attack, we can say that the vulnerability is high. There is a high probability that the patient will die.

Vulnerability is an aspect of risk. Because of this, the vulnerability analysis is a part of the risk analysis. If vulnerability is highlighted in the analysis, we often talk about risk and vulnerability analyses.

Resilience is closely related to the concept of robustness. The key difference is the initiating event A. Robustness and vulnerability relate to the consequences and probabilities given a fixed A, whereas resilience is open for any type of A, including surprising events. We may get ill due to different types of virus attacks; also new types of viruses may be created. From this idea we define resilience as (Steen and Aven, 2011):

Resilience: $(C, P_f \mid \text{any A, including new types of A})$

and the resilience description:

$(C, P_f{}^*, U(P_f{}^*), K \mid \text{any A, including new types of A})$.

Hence the resilience is considered high if the person has a low frequentist probability of dying due to any type of virus attack, also including new types of viruses. Resilience is about the consequences in the case of any "attack" (virus attack) and associated probabilities. We say that the system is resilient if the resilience is considered high. Of course, in practice we always have to define some boundaries for which A events to allow for.

For all these definitions, the consequences C depend on the performance of barriers (denoted B) (Flage and Aven, 2009), and to explicitly show this we write $C = (B, C)$, resulting in a vulnerability definition $(B, C, P_f \mid A)$, etc.

The performance of the barrier can be expressed through the capacity of the barrier (and associated probability), for example the strength of a wall. The barriers and the system performance in general are affected by a number of performance-influencing factors (PIFs), for example resources, level of competence, management attitude, etc.

Analogous to risk assessment and risk management we define vulnerability assessment, vulnerability management, resilience assessment and resilience management (engineering), for example:

Resilience engineering is all measures and activities carried out to manage resilience (normally increase resilience).

The above set-up is motivated from the belief that it provides a logically defined structure for risk, vulnerability and resilience. But do we need all these concepts and terms? It is appropriate to question whether too many definitions could be a hindrance to professional practice and/or intellectual discourse, as discussed in Aven (2011). One suggestion for simplifying the terminology is to completely remove the concept of resilience as indicated by Aven (2011). Vulnerability (antonym robustness) is then the term used for reflecting "risk" conditional on the occurrence of one or a set of events A. It is not distinguished between whether the events are known or unknown. The point is that we always have to define the set of events that the "risk" is conditional on when talking about robustness/vulnerability. A number of indices can be defined to measure vulnerability/robustness but we do not need names for all types of such indices. Resilience is then captured by the concept of vulnerability/robustness. This comment also applies to the set-up in the next section. For the analysis in this book, the distinction between vulnerability/robustness and resilience is, however, not so important as these terms will not often be used.

2.8 A risk assessment perspective based on the (A,C,U) definition

Then we come to the second risk perspective that we will look closer at in the coming chapters, the (A,C,U)-based perspective.

By replacing the relative frequency-interpreted probability P_f by uncertainty U, we obtain the (A,C,U) risk perspective, where risk is the two dimensional combination of

 (i) events A and their consequences C, and
(ii) the associated uncertainties U about A and C (will A occur and what will the consequences C be?), including uncertainty about underlying phenomena influencing A and C.

Often the A events are specified, for example as gas leakages in a process plant or as terrorist attacks in a country, but we may also allow for new types of such events, a new type of virus for instance. We speak then often about "unknown uncertainties" ("unknown unknowns", ignorance or non-knowledge) – we do not know what we do not know, in contrast to "known uncertainties" ("known unknowns") – we know what we do not know (Aven and Renn, 2009b).

A risk description based on this definition would cover the following components:

$$\text{Risk description} = (A,C,U,P,K), \tag{2.1}$$

that is, risk is described by events A and consequences C, knowledge-based probabilities P, uncertainties U not captured by P, and K the background knowledge that U and P are based on. This description covers probability distributions of A and C, as well as predictions of A and C, for example a predictor C* given by the expected value of C, unconditionally or conditional on the occurrence of A, i.e. $C* = EC$ or $C* = E[C|A]$. The U in (2.1) may for example be a qualitative assessment of uncertainty factors (assumptions on which the probabilities are based); see the coming section (and Chapter 6). It partly relates to the unknown unknowns.

As subjective probabilities are used, this perspective is also Bayesian, although it has a different focus and is based on other building blocks than the probability of frequency approach studied in Section 2.4.

Also in this perspective we may introduce probability models (with parameters) – chance models – expressing aleatory uncertainty, i.e. variation in populations of similar units. However, such models need to be justified, and if introduced they are to be considered as tools for assessing the uncertainties about A and C. The assessment of the parameters of the models is not the end product of the analysis as in the (A,C,P_f) approach. The parameters of the model are in this case treated as unknown quantities, as C.

Vulnerability and resilience

As in the previous section we may introduce the concepts vulnerability and resilience (Aven, 2008a; Steen and Aven, 2011):

$$\text{Vulnerability (antonym robustness)} = (C,U \mid A),$$

in other words, the vulnerability is the two-dimensional combination of consequences C and associated uncertainties U, given the occurrence of an initiating event A. The related description of vulnerability thus covers the following elements:

$$(C,U,P,K \mid A)$$

i.e. the possible consequences C, uncertainty U, probability P and the background knowledge K, given that the initiating event A takes place. In line with Aven and Renn (2009a), we may interpret vulnerability in relation to the event A as uncertainty about and severity of the consequences of an activity given the occurrence of A.

Similarly we define resilience and resilience description:

$$\text{Resilience: } (C,U \mid \text{any A, including new types of A})$$

$$\text{Resilience description: } (C,U,P,K \mid \text{any A, including new types of A}).$$

The performance of barriers B can be included explicitly as explained in the previous section, for example leading to a vulnerability description $(B,C,U,P,K \mid A)$.

2.9 Example: Offshore diving activities

Let us go back to the 1970s and consider the risk related to future health problems for divers working on offshore petroleum projects at that time. We distinguish between the two risk definitions and descriptions introduced in Sections 2.7 and 2.8, i.e. the (A,C,P_f)-based perspective and the (A,C,U)-based perspective. Let us first look at the probability of frequency approach. A relative frequency-interpreted probability P_f is defined expressing the probability that a random chosen diver would experience health problems (properly defined) during the coming 30 years due to diving activities. The probability P_f is unknown and needs to be estimated. Based on the available knowledge at that time, an estimate $P_f^* = 0.01$ of P_f is established. Uncertainties are expressed using subjective probabilities according to the probability of frequency approach. A 90 per cent credibility interval [0.001, 0.1] is computed, expressing that the posterior probability that P_f lies in this interval is 90 per cent given the available data and knowledge about the phenomena in general. There are not strong indications that the divers will experience health problems. This would be a risk description $(A,C,P_f^*,P(P_f),K)$ according to the probability of frequency approach. It is a standard risk assessment description which includes second-order probabilities, in line with the Bayesian approach. Refer to Appendix A.2.2 for explanations of the Bayesian terminology used (credibility interval, posterior probability).

These probabilities indicate that one would not expect severe health problems for the divers in the future. However, we know today that these probabilities led to poor predictions. A large number of the divers have experienced severe health problems (Aven and Vinnem, 2007, p. 7). Also at the time of the assessment there were uncertainties about the future health conditions, but the uncertainties were not revealed by the probabilistic analysis.

To improve the assessment, a change of the main risk index is suggested: instead of P_f, focus is on the probability distribution F_f of the proportion D of divers that will experience health problems. Note that P_f equals the expected value of D. To see this let n be the number of divers in the population, and let I(B) denote this indicator function for the event B, which is equal to 1 if B occurs and 0 otherwise. Then we see that

$$P_f = \Sigma_j \, P \text{ (person j experiences health problems)}/n$$
$$= E \, \Sigma_j \, I \text{ (person j experiences health problems)}/n = ED,$$

which proves the assertion.

The assessment produces for example an estimate $p^*(0.1)$ of the probability $p(0.1)$ that more than 10 per cent of the divers would get health problems. We may also express second-order probabilities for example using credibility intervals as shown above for P_f. Suppose that $p^*(0.1) = 0.05$ and a 90 per cent credibility interval equals $[0.001, 0.20]$.

Certainly, this change of risk parameter, seeing beyond the expected value-based parameter P_f, has given a more informative risk description, but still the conclusion would be that we do not expect severe health problems for the divers in the future; the probabilities are low.

Next we consider the problem following the (A,C,U) risk perspective. Relative frequency-interpreted probabilities are not introduced. The risk description covers (A,C,U,P,K) using the above notation. The assessment covers a probabilistic analysis producing for example an assigned probability distribution of D, a 90 per cent prediction interval $[a,b]$ of D (such that $P(a \le D \le b) = 0.90$), as well as an assignment of the probability that a random chosen diver would experience health problems. In addition, assessment of uncertainties beyond probabilities is required.

This uncertainty assessment could for example take the following form. First a set of uncertainty factors are identified. These factors relate to the underlying understanding of relevant physiological and psychological phenomena, as well as assumptions and suppositions made in the probabilistic analysis. An example of such a factor is the diving operation's effect on the brain, in particular related to long-term effects on forgetfulness. The data used as a basis for the analysis do not show a significant difference between divers and non-divers. However, the data material is not extensive and is limited to a rather short-time interval compared to the 30 years addressed in the risk assessment.

Each factor's importance is measured using a sensitivity analysis. Is changing the factor important for the risk indices considered, for example the distribution of D? If this is the case, we next address the uncertainty of this factor. Are there large uncertainties about this factor? If the uncertainties are assessed as high, the factor is given a high risk score. Hence, to obtain a high score in this system, the factor must be judged as important for the risk indices considered and the factor must be subject to large uncertainties.

One may question whether it is not possible to include the uncertainty factor explicitly in the probabilistic calculations. For some factors that could be possible, in some cases. Consider for example the above factor related to long-term effects on forgetfulness. Suppose the probabilistic assessment was first based on the assumption that the diving had no effect on the forgetfulness. Then we can perform a reassessment where this assumption is left out.

The result would be small changes in the probabilities, as the assessment would be dominated by the view that there is no information supporting the rejection of the assumption. The importance of the uncertainty is marginal measured by the probabilities.

By addressing the issue as an uncertainty factor its importance is increased. For the decision-makers who need to balance different concerns and determine to what extent principles of caution and precaution should be applied, such information is required.

In this particular case, it is obvious that weight given to the cautionary and precautionary principles would have significantly reduced or even ended the diving operations at that time. However, the task of balancing different concerns and giving weight to the uncertainties is a management (here political) responsibility, and the decision made was to perform such diving operations. We may just speculate whether a risk perspective highlighting the uncertainties would have changed the decision. It could have, but probably not as the economic incentives for performing the activities were so strong.

2.10 Summary of concepts and perspectives

We have described two main categories of risk perspectives, the (A,C,P_f) and (A,C,U)-based perspectives, with associated risk descriptions:

$$\text{Risk description in the } (A,C,P_f) \text{ case} = (A,C,P_f{}^*,U(P_f{}^*),K),$$
$$\text{Risk description in the } (A,C,U) \text{ case} = (A,C,U,P,K).$$

For the (A,C,P_f) case we have considered two ways of expressing the uncertainties $U(P_f{}^*)$, using knowledge-based (subjective) probabilities which leads to the probability of frequency approach, and using confidence intervals based on relative frequency-interpreted probabilities and hard data only (pure statistical approach).

The aim of the probability of frequency approach is to describe the uncertainties of the underlying frequencies P_f and we may consider it a special case of the (A,C,U,P,K) description. Consequently we can restrict attention to two main categories of perspectives for the risk assessments (QRAs): the pure statistical approach where we seek accurate estimates of the underlying true risk (relative frequency-interpreted probabilities) and an approach where the aim of the risk assessment is to describe the uncertainties about the unknown quantities of interest (represent/express the knowledge and lack of knowledge available). These two perspectives will be studied in Chapters 5 and 6, linked to the three cases introduced in Chapter 4.

3

Science and scientific requirements

In this chapter we present the announced requirements of reliability and validity that will be used to verify that a risk assessment is scientific (Section 3.3). But first in Section 3.1 we give some reflections about risk assessment being a scientific method motivated by two interesting editorials of the first issue of the journal *Risk Analysis* (Cumming, 1981; Weinberg, 1981), in relation to the establishment of the Society of Risk Analysis. We also provide a brief review of the traditional sciences (Section 3.2), such as the natural sciences, social sciences, mathematics and probability theory, to place risk assessment into a broader scientific context. A key issue is to what extent risk assessment should be judged by reference to these traditional science paradigms, or is a science per se.

3.1 Reflections on risk assessment being a scientific method

Cumming (1981) concludes that the process of analysing or assessing risks involves science, and consequently is a scientific activity. However, according to Cumming, risk assessment is not a scientific method per se. He writes:

Risk assessment cannot demand the certainty and completeness of science. It must produce answers because decisions will be made, with or without its input. The quality of societal decisions will be influenced by the quality of the risk information which goes into them, and the long term success of a society is influenced by the quality of its decisions. Thus, risk assessment is an important activity. It depends on science and has an important stake in receiving the input of good science.

Cumming sees some useful functions of the society and the new journal, but also some dangers:

The formation of the Society may imply to some that risk assessment is indeed a "science" and lead them to expect a degree of precision which is not now possible.

Professionalization may tend to hide or disguise the nature, the end uses, and the trans-scientific elements of risk assessment, and to confer upon it the image of power and sophistication it does not have.

By the trans-scientific elements of risk assessments he refers to predictions of rare events where the uncertainties are very large, for example the number of deleterious biological effects resulting from exposure to environmental insults at dose levels far below the levels at which effect can be seen (Weinberg, 1972).

Weinberg in his editorial states some similar comments:

I welcome the new journal, *Risk Analysis*, where practitioners of the art of risk assessment can exchange their views and results. I use the word "art" intentionally: I can hardly conceive of large parts of risk assessment becoming a science. This is not to say that careful analysis of underlying assumptions that go into risk assessment is fruitless; or that careful observation of damage caused by large insults is not a part of science. But there are, and will always be, strong trans-scientific elements in risk assessment. We should be prepared to recognize these, and accept them.

Weinberg (1981) stresses that the most powerful method of science – experimental observation – is inapplicable to the estimation of overall risk in the case of rare events, which are those instances where public policy most often demands assessment of risk. He refers to the intrinsic uncertainty reflected in the bitter controversy that rages over the safety of nuclear reactors.

We observe that both editorials conclude that risk assessment comprises scientific elements but is not a scientific method per se. The main problem seems to be the lack of predictability in the case of large uncertainties. Their reference is obviously the "traditional scientific method" which is the pillar for the natural sciences. The method is based on the collection of data through observation and experimentation, and the formulation and testing of hypotheses. More specifically it is common to distinguish between four steps of the method (Wolfs, 2009):

1. observations and descriptions of a phenomenon
2. formulation of a hypothesis to explain the phenomena, for example expressed by a mathematical formula
3. use of the hypothesis to predict the results of new observations
4. performance of experimental tests to verify or falsify the hypotheses.

If the experiments bear out the hypothesis it may come to be regarded as a theory or law of nature. If the experiments do not bear out the hypothesis, it must be rejected or modified. What is key in the description of the scientific method just given is the predictive power (the ability to get more out of the theory than you put in) of the hypothesis or theory, as tested by experiment.

In practice the scientific method does not always follow these steps. For example, the process may not start with observations. A new hypothesis may be inspired by reading what others have done and by discussions with colleagues. Many scientific investigations use a set of methods, including experimentation, comparison and modelling. The results from one research study may lead in directions not originally anticipated, or even in multiple directions as different scientists pursue areas of interest to them (Carpi and Egger, 2003).

But regardless of how the scientific method is carried out, its aim is accurate predictions. Cumming (1981) and Weinberg (1981) quickly conclude that risk assessments are not able to meet this ambition: accurate predictions cannot be made in case of rare events and large uncertainties. However, in many cases the scientific method as defined above would be applicable and it will be useful to clarify the boundaries of risk assessment as a tool for this purpose. This is a main objective of Chapter 5. More important though is the question whether this is an appropriate aim of risk assessment. Seeing risk assessment as a tool for expressing or measuring uncertainties about unknown quantities would obviously not be in line with the scientific method as defined above, but could it still be seen as science? This issue we will discuss in Chapter 6. As a background for this discussion we need to rethink what science means and what are reasonable requirements to be set to a scientific method. But first we give a review of some of the traditional sciences, starting with the natural sciences which are based on the scientific method outlined above. Some people may consider only the natural sciences as "true sciences", but in our view such a perspective is meaningless. Yes, the scientific method as described above has been very successful in natural sciences, but are not mathematics and probability theory science? Of course they are. There exist a number of scientific journals covering these disciplines and there exist many recognised scientists in these fields. Are these journals and these scientists second-class representatives for science? No, these disciplines are not comparable with the scientific method and the natural sciences, they could be used in these sciences and when applying the scientific method, but the reference for making a judgment about being scientific is not the natural sciences and the scientific method alone. We will explain this in more detail in the coming section.

In this book we are concerned about risk assessment as a scientific method (in a broad sense, not only according to the traditional scientific method outlined above). Risk assessment as a scientific method is not the same as risk assessment as a scientific discipline. We may view the science of risk assessment as the development of concepts, principles, methods and models to analyse and evaluate (assess) risk, in a decision-making context (Aven, 2004).

This science, which is founded on the international scientific journals in the field, provides the basis for the scientific method of risk assessment, provided that such a method can be justified.

When interpreting Cumming (1981) and Weinberg (1981), one has to be aware that risk analysis is often defined in different ways as was noted in Chapter 1. Our definition (refer to Section 1.1) states that risk analysis is a methodology designed to determine the nature and extent of risk, comprising the following three main steps:

- identification of hazards/threats/opportunities (sources)
- cause and consequence analysis, including analysis of vulnerabilities
- risk description, using probabilities and expected values.

The most common alternative definition, often used among members of the Society of Risk Analysis, sees risk analysis as an overall concept, comprising risk assessment, risk perception, risk management, risk communication and their interaction (refer to IRGC 2005). However, Cumming (1981) and Weinberg (1981) in their reflections are precise in referring to the use of risk assessments and not risk analysis, so these different uses of the term risk analysis should not be a problem.

3.2 Review of some traditional sciences important for risk assessment and risk management

Natural sciences

The natural sciences provide theories and laws describing the physical world. A theory is often defined as a set of statements or principles devised to explain a group of facts or phenomena, especially one that has been repeatedly tested or is widely accepted and can be used to make predictions about natural phenomena, whereas a law in this context is often defined as a property of a physical phenomenon, or a relationship between the various quantities or qualities which may be used to describe the phenomenon, that applies to all members of a broad class of such phenomena, without exception. An example of a physical law is Newton's second law of motion, $F = ma$.

These theories and laws are established primarily based on theoretical reasoning, and statistical data analysis is used to refine or reject the laws and theories. Developing theories and laws is a creative process that is difficult to formalise and document. Theories and laws cannot be proved true. Observations may to varying degrees support the theories and laws, and this results in new assumptions and revision of the theories and laws. Rejection of the theories and laws may also be the result in some cases, but, normally, only if alternatives exist.

Theories and laws can be used to define models of the world, i.e. simplified representations of the world for specific situations. A model could be more or less good at describing the world (it is always wrong, otherwise it would not be a model – a simplification of the world) but still it could be useful for its purpose.

An example of a physical model is the expression for speed v of an object dropped from a height h, derived by assuming that the kinetic energy of the object at the reference point equals the potential energy at h:

$$v = \sqrt{2gh}, \tag{3.1}$$

where g is the acceleration due to gravity (Nilsen and Aven, 2003). This is a good model for describing the world in many cases. But it is a model, a simplified representation of the world, as it disregards the air resistance, and that g varies through the fall. Consider an application of the model given by Equation (3.1) in order to predict the velocity of an object dropped from a crane located on a floating structure. It can be argued that the vertical motion of the structure due to ocean waves would cause the model to be inaccurate. Such motion would affect the height of fall h, cause an initial speed $v_0 \neq 0$, and a relative vertical motion of the object hit by the dropped object, i.e. three effects that are not taken into account by the model. Despite these inaccuracies, we could find this simple model to be a good model for describing the world in a specific case.

Mathematics and probability theory

The science of mathematics and probability theory is based on some axioms, and deduction from these using the rules of logic. For example, probability theory is normally based on three axioms: a probability is a non-negative number, the probability of a certain event is 1, and the probability of a union of mutually exclusive events is equal to the sum of the probabilities of each event. In addition we need to define a conditional probability. From these axioms, we derive the many well-known probability rules. Accepting the rules of logic, mathematics and probability theory provide 100 per cent certain knowledge, but no information about the world per se.

Traditional statistical inference

Traditional statistical inference is based on the following thinking: observations are sampled from a large population of units. Variations within the population and sample are described using a stochastic model; and combining the data and the model, conclusions are made on performance measures related to the whole population, or specific units of the population.

Bayesian theory

Traditional Bayesian theory is based on the use of subjective probabilities to express uncertainties about unknown quantities, both observable quantities of the world and parameters of models of the world. Probability theory is used to manipulate the probabilities. The probabilities are updated when new information becomes available in a coherent way using Bayes' formula.

Bayesian decision analysis includes, in addition to this, specification of utilities for possible outcomes of decision alternatives, and the use of the maximum expected utility as a decision criterion.

Social sciences

A main line of scientific reasoning within the social sciences is based on the application of traditional statistical inference, as described above. More specifically, the method covers the formulation of a hypothesis, theoretical definition of this hypothesis (using probabilistic terms, like parameters of a distribution function), operational definition (using observables), statistical testing and then conclusions and discussions.

Theories involving human beings and society are different from theories in natural sciences as the former theories become a part of the human beings and society and influence their behaviour and development. When analysing human beings and society there is a need for a *double reflection*, to explain how the theory would affect human beings and society. This leads to a theory of the theory, and this can be continued infinitely.

Other sciences

There are a number of other sciences that are relevant for risk analysis, such as technology, medicine, psychology, etc., but these will not be further discussed in this book. For the purpose of the present analysis the above sciences provide a sufficiently broad spectrum of reference sciences to discuss the foundation of the science of risk assessment and risk management.

All these sciences can be seen as sciences per se, but also as tools for the scientific method defined above. To analyse risk we use physical models and models for describing human and organisational aspects, for example the model (3.1). Mathematics and probability, traditional statistical inference and Bayesian analysis are key tools for implementing the scientific method. There are different views on which tools should be preferred, for example Bayesian analysts would favour a completely different approach than proponents of the traditional statistical inference. For example, Lindley (2006) argues that the Bayesian perspective provides an overall suitable framework for the scientific method, as this perspective allows for descriptions of uncertainties and systematically incorporates new knowledge. Others would reject the Bayesian perspective as it fails to produce "objective results". This discussion is also relevant for risk assessment and will be thoroughly analysed in the coming section and chapters of the book.

Next we will discuss what general requirements should be set to risk assessment in order for it to be a scientific method.

3.3 Risk assessment as a scientific method. The reliability and validity requirements

For risk assessment to be a scientific method the following requirements should be met (Aven and Heide, 2009):

1. The scientific work shall be in compliance with all rules, assumptions, limitations or constraints introduced, and the basis for all choices, judgements etc. given shall be clear, and finally the principles, methods and models shall be subjected to order and system, to ensure that critique can be raised and that it is comprehensible.
2. The analysis is *relevant and useful* – it contributes to a development within the disciplines it concerns, and it is useful with a view to solving the "problem(s)" it concerns or with a view to further development in order to solve the "problem(s)" it concerns.
3. The analysis and results are *reliable* and *valid*.

Requirements 1–2 are based on standard requirements for scientific work (RCN, 2000). The purpose of risk assessment is to provide decision support by systematisation of knowledge to describe/express risk. This is a unique objective for risk assessment. Terminologies have been developed, as well as principles and methods for assessing risks. However, there is no broad consensus in the risk assessment community about the terminology and the principles and methods to be used. For example, risk is defined and expressed in many different ways. The full discussion of these issues is beyond the scope of this book, but important aspects are covered by our analysis in Chapter 2 and in the rest of this book, in particular in relation to our discussion of requirement 3, the reliability and validity criteria. These relate in particular to how to understand and express risk, as well as the use of the risk assessments.

The definitions of the terms "reliability" and "validity" in a risk assessment context are not obvious, but a logic and structure for interpreting these concepts have been established by Aven and Heide (2009) and this will be used in this book.

While reliability is concerned with the consistency of the "measuring instrument" (analysts, experts, methods, procedures), validity is concerned with the success at "measuring" what one sets out to "measure" in the analysis. More precisely we make the following definitions:

Reliability: The extent to which the risk analysis yields the same results when repeating the analysis (R).

Validity: The degree to which the risk analysis describes the specific concepts that one is attempting to describe (V).

Depending on the objectives of the analyses, more specific and detailed interpretations (sub-criteria) of the above general definitions of reliability and validity can be formulated (Aven and Heide, 2009):

Reliability

The degree to which the risk analysis methods produce the same results at reruns of these methods (R1).

The degree to which the risk analysis produces identical results when conducted by different analysis teams, but using the same methods and data (R2).

The degree to which the risk analysis produces identical results when conducted by different analysis teams with the same analysis scope and objectives, but no restrictions on methods and data (R3).

Validity

The degree to which the produced risk numbers are accurate compared to the underlying true risk (V1).

The degree to which the assigned probabilities adequately describe the assessor's uncertainties of the unknown quantities considered (V2).

The degree to which the epistemic uncertainty assessments are complete (V3).

The degree to which the analysis addresses the right quantities (V4).

In the following two chapters we will discuss to what extent these requirements are met, by using the three cases as illustrating examples.

4

Introduction to case studies

In this chapter we introduce the three case studies which will be pursued through the rest of the book to illustrate concepts, principles and methods. The first of these studies is related to the analysis of working accident data whereas the second relates to the risk assessment of a Liquefied Natural Gas (LNG) plant and the third relates to a design of a safety system. Our starting point for the LNG case is the risk assessment process as carried out recently for the plant (Vinnem, 2010). We have, however, made some adjustments, to be in line with the principles studied in this book. For the purpose of this book we consider different approaches for how to carry out the assessments. The presentation only shows excerpts from the assessment; it is simplified, and many numbers have been changed.

4.1 Working accidents

In 1999 the Petroleum Safety Authority Norway (PSA) took the initiative to develop a method in order to assess status and trends for the risk levels in the Norwegian offshore petroleum industry (Vinnem *et al.*, 2006). A method was developed which is based on recording occurrences of near misses and relevant incidents, performance of barriers and results from risk assessments, as well as evaluation of safety culture, motivation, communication and perceived risk. In this book we focus on the analysis related to occupational accidents. Accident statistics are provided for serious injuries, which are defined by PSA (2007) as:

- head injuries involving concussion, loss of consciousness or other serious consequences,
- loss of consciousness as a result of working environment factors,

Number of serious injuries per million manhours

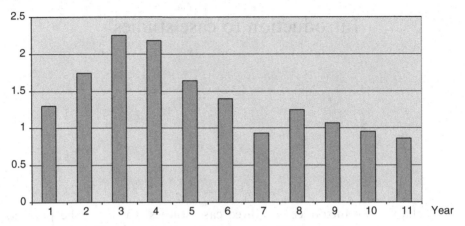

Figure 4.1   Number of serious injuries per million manhours.

- skeletal injuries, with the exception of simple hairline fractures or fractures of fingers or toes,
- injuries to internal organs,
- whole or partial amputation of parts of the body,
- poisoning with danger of permanent health injury, such as H_2S poisoning,
- burns, frost or corrosive injuries with full thickness skin injury (third degree) or partial thickness skin injury (second degree) to the face, hands, feet or abdomen, as well as all partial thickness skin injury that covers more than five per cent of the surface of the body,
- general cooling (hypothermia),
- permanent, or for a long period of time, inability to work.

The number of serious injuries per million working hours for the previous 11 years is presented in Figure 4.1. The numbers are based on about 30–40 million working hours per year, and the number of injuries ranging from about 70 to 30. The total number of installations is close to 100.

The data are split into production units and mobile units, as well as key functions: administration and production, drilling and well operations, catering, and construction and maintenance. Figures 4.2 and 4.3 show the injury frequencies for fixed installations and mobile units, and the data for the construction and maintenance function separately.

Figure 4.2 shows that the injury frequency has a falling trend from year 3 to year 11. From year 10 to 11 there was a change in frequency from 0.9 in year 10 to 0.7 in year 11. On the production (fixed) installations there were 19 cases of serious injury in year 11. The injury data for the construction and

Number of serious injuries per million
manhours. Production installations

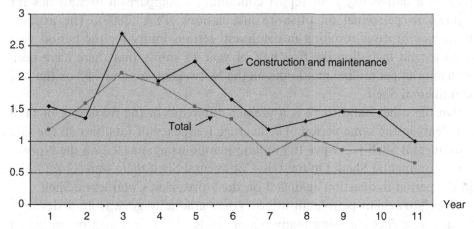

Figure 4.2   Number of serious injuries per million manhours. Production
(fixed) installations. For the function construction and maintenance, and the
total for all functions.

Number of serious injuries per million manhours.
Mobile units

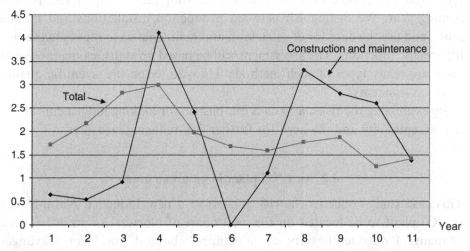

Figure 4.3   Number of serious injuries per million manhours. Mobile units.
For function construction and maintenance, and the total for all functions.

maintenance function shows a similar form. The frequency for mobile units
follows the same type of pattern, but the data for construction and maintenance show more variation. In year 11 there was an increase in the total injury
frequency compared to the previous year, from 1.3 in year 10 to 1.4 in year 11.

PSA Norway and the Health and Safety Executive (HSE) in the UK publish a half-yearly joint report containing a comparison of statistics for injuries to personnel on offshore installations (PSA, 2009). The average frequency of cases resulting in death and serious injury for the period year 4 up to and including the first half of year 11 shows that there have been 0.96 injuries per million manhours on the Norwegian side and 1.03 on the UK Continental Shelf.

On the other hand there is a greater difference in the frequency of fatal accidents in the same period. The average frequency of fatalities on the UK Continental Shelf is 2.9 per 100 million manhours against 1.2 on the Norwegian Continental Shelf. On the UK Continental Shelf there were 11 fatalities in the period in question against 3 on the Norwegian Continental Shelf.

The Norwegian petroleum industry has gradually gone from a development phase encompassing many major fields to one in which operation of facilities dominates. Measuring the effect of the total safety work in these activities is considered a key challenge for the authorities, and the aim of the developed method for analysing the risk level has been to contribute to this end. In this book we address the following question:

Based on the statistics available on serious injuries, we perform a quantitative risk assessment with the aim of expressing serious injury risk for the coming year. We distinguish between production installations and mobile units, and use the data for construction and maintenance to illustrate the risk for specific functions. The assessment will be based on statistical methods but there are many types of such methods. How do we ensure scientific quality of the assessments?

By expressing the risks, a basis is established for making judgements about risk acceptability and the need for further risk-reducing measures.

4.2 An LNG plant in an urban area

This case study concerns the risk related to a new Liquefied Natural Gas (LNG) plant to be located on the west coast of Norway, in an urban area (Tananger) outside the city of Stavanger, about 4 km from Stavanger Airport. Despite the formal approval according to the SEVESO II Directive, there is considerable resistance to the plant from the neighbours living less than one kilometre from the plant. The LNG plant is located only a few hundred metres from a ferry terminal and this also creates concern.

The LNG plant is now under construction by the energy supplier Lyse. The necessary approval from local and central authorities has been obtained. The plan is that natural gas from the North Sea is transported through pipelines to shore, and then liquefied at the plant before being stored in a huge tank. The LNG is then distributed from the plant to local consumers by LNG tankers and LNG lorries (Vatn, 2010). The annual production is 300 000 tons of LNG, but the capacity may be increased to 600 000 tons if market conditions allow such an increase.

The LNG plant has the following main components (Vinnem, 2010): Pipeline landfall, Gas reception facilities, Pre treatment, LNG production, LNG tank and Export facilities.

The following risk studies have been carried out during the planning and engineering process:

1. Several Preliminary Hazard Analyses (PHA)
2. Initial QRA study (Lyse, 2007)
3. QRA study in detailed engineering (Lyse, 2008).

The PHA studies were essentially conducted by a Norwegian consultancy, whereas the initial QRA study was conducted by a UK consultancy. The QRA during detailed engineering was conducted by the engineering contractor.

The operator Lyse has adopted a traditional risk assessment approach based on steps as in Figure 1.2. Figure 4.4 shows the steps adopted for the QRA.

The risk assessments produce probabilities and these are compared to a set of risk acceptance criteria. These apply to first, second and third parties, defined as:

• The first party risk is defined as a fatality risk for the Lyse LNG base load plant personnel.
• A fatality risk for the LNG Carrier personnel (truck, ship loading and external contractors) is defined as second party risk.
• Third party risk covers people living, working or staying outside the Lyse LNG base load plant.

In this case we focus on the third party risk.

Two categories of risk acceptance criteria are defined for third party risks: individual risk (IR) based criterion and a frequency-number (f-n)-curve based criterion. Both criteria define an upper unacceptable (intolerable) risk level and a lower level of acceptable risk. Between these limits the ALARP principle applies. The IR is defined as the probability that a specific person (arbitrarily chosen) shall be killed due to the activity during a period of one year. The f-n curve represents the frequency f (i.e. the expected number) of accidents that lead to n or more fatalities, which is approximately equal to the probability of an accident with at least n fatalities.

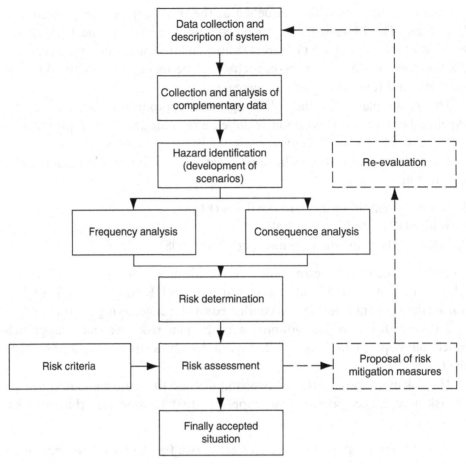

Figure 4.4　The QRA process adopted in Lyse (2008).

IR > 10^{-5}	Unacceptable risk
10^{-5} < IR < 10^{-7}	ALARP
IR < 10^{-7}	Acceptable risk

Figure 4.5　The risk acceptance criteria adopted in the case based on IR and 3rd party risk.

The following limits were defined for IR; see Figure 4.5:

$$IR > 1 \times 10^{-5} \quad \text{Unacceptable (intolerable)}$$
$$IR < 1 \times 10^{-7} \quad \text{Acceptable risk}$$

In the region (10^{-7}–10^{-5}) the ALARP principle is applied.

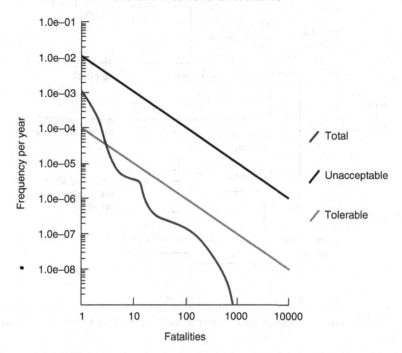

Figure 4.6 Risk acceptance criteria (unacceptable risk level, acceptable risk level and ALARP region) and calculated f-n curve (Lyse, 2007).

The limits for the f-n curve are shown in Figure 4.6, together with the calculated results in the Lyse (2007) report.

Based on Figure 4.6, Lyse concluded that the risk level was negligible. Also an independent research team which was hired by the plant developer in order to improve risk communication commented that the risk was considered to be negligible (Vatn *et al.*, 2008). Also the IR numbers were below the unacceptable risk level.

The risk assessment process has given rise to a substantial amount of professional discussion (Vinnem, 2010). The main issues have been the quality of the risk assessments, and in particular the use of pre-defined risk acceptance criteria and how to treat uncertainties. These issues will be thoroughly studied in this book. The main question we ask is to what extent the risk assessment approach adopted meets the standards that we can expect from a scientifically based assessment. Specifically we address the following points:

- The rationale for defining risk acceptance criteria, reflecting that risk in general needs to be balanced against benefits, and the lack of precision

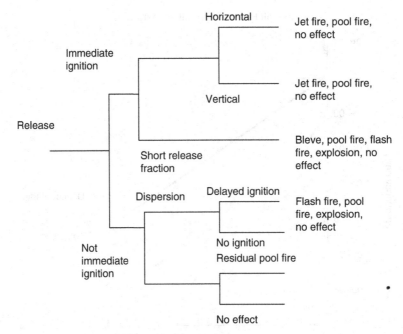

Figure 4.7   An example of an event tree used in the QRA (based on Lyse, 2008).

of risk assessment to perform direct comparisons between risk results and the criteria.

• How to determine the specific levels for unacceptability and acceptability.

The risk assessments performed were based on a number of assumptions, and these strongly affect the results of the assessments. How can we then rely on the results of the assessments? In the assessments we use a number of models and procedures for calculating the risk. How should we understand the concepts of model inaccuracy and model uncertainty? A model, like for example an event tree as shown in Figure 4.7, is obviously a simplification of the real world, and it could produce inaccurate predictions. The event tree shows for example that if a hydrocarbon release occurs, the outcome could be a fire, bleve, explosion, etc. depending on the result of the branching events. The real world is more complicated. But to what extent should the inaccuracies be reflected in the assessments? Should we quantify the inaccuracies? In the case study this issue was not addressed at all. Model inaccuracy (uncertainty) was not discussed in the risk assessments report. In fact uncertainty as a concept was not mentioned at all in the reports. How is that possible in a QRA which to a large extent is about uncertainties and quantification of uncertainties? The scientific quality of the assessment certainly needs to be discussed.

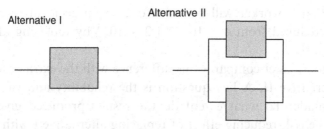

Figure 4.8   The two system alternatives considered for the design of a safety system.

4.3 The design of a safety system

A safety system is to be designed in a process plant. Two alternatives are considered: I and II as shown in Figure 4.8.

Alternative I comprises only one unit whereas alternative II is a redundant system with two units in parallel. Hence, for alternative II a system failure occurs only if both units fail. Alternative II is expected to have a higher reliability than alternative I; however, an assessment is required to "measure" the difference. The decision-maker is not convinced that the increased safety obtained by implementing alternative II can be justified given the additional costs that are associated with this alternative.

The analysts performing the assessment develop models of the two system configurations. The models are based on the assumption that the units are periodically tested after time intervals of length τ. If it is discovered that a unit has failed, it is repaired (replaced) and it is then assumed to be as good as new, and the process repeats. The state of the units can only be revealed by testing. The times that the units are being tested and repaired are ignored.

Some data on the failure frequency exist. The analysts estimate that the units have a failure rate λ equal to 0.2 per year, and based on this figure they compute a mean fractional dead time (MFDT) by the formulae (Aven, 1986; see also Section 5.4):

$$\text{Alternative I: } \lambda\tau/2$$
$$\text{Alternative II: } (\lambda\tau)^2/3.$$

We interpret the MFDT as the fraction of time the system is not functioning in the long run.

For $\tau = 1$ this gives a MFDT equal to 0.1 (10%) for alternative I and 0.013 (1.3%) for alternative II.

The difference is large but somewhat reduced when considering the influence on the risk for the total plant. For example the individual risk (the probability that a worker will be killed due to an accident in a period of one year) is reduced from 4×10^{-4} to 2×10^{-4} by replacing alternative I by alternative II.

The decision-maker compares this difference with the extra costs of implementing alternative II. A key question is the reliability and validity of the assessment made. To what extent do the results produced give the "true picture" of the risk-reducing effect of replacing alternative I with alternative II? Furthermore, the assessment is based on models and a set of assumptions. How should the possible model inaccuracy be taken into account? Is the model sufficiently accurate for its purpose? There is a need to discuss the scientific quality of the assessment.

5

Risk assessment when the objective is accurate risk estimation

In this chapter we study the scientific platform of risk assessments when the objective of these assessments is accurate risk estimation. We first summarise the key concepts, probability and risk, using the set-up introduced in Chapter 2. We then conduct the assessments for the three cases, and from this basis we study the scientific quality of the risk assessments. Focus is on the scientific requirements of reliability and validity defined in Chapter 3.

The risk assessments presented in Sections 5.2–5.4 are rather comprehensive and detailed, although many simplifications have been made. If the reader is mainly concerned about the discussion of the scientific requirements of reliability and validity, a quick reading of these sections would suffice provided the reader is familiar with the statistical nomenclature and methods used. However, to fully appreciate the discussion in Section 5.5 it is necessary to go into the details of the assessments in Sections 5.2–5.4. For example, we cannot evaluate what the main quantities of interest are in the study or see the importance of key assumptions made in the assessments, without looking into the contexts of the analyses and precisely describing how the analyses are carried out.

5.1 Scientific basis

We consider an activity and make the following definition of risk associated with this activity:

Risk = (A,C,P_f), where P_f is a relative frequency-interpreted probability (or a related parameter such as the expected number of occurrences of the event A per unit of time, where expectation is with respect to a relative frequency-interpreted probability).

As an example consider the operation of a plant and the risk associated with accidents that could lead to fatalities. Then we can define events (such as gas leakages), associated consequences (losses) and probabilities expressing for example the probability P_f of at least 10 fatalities next year. The P_f is interpreted as the fraction of years where this event occurs in an infinitely large population of similar plant years, as will be discussed in more detail in the coming section.

According to this definition, risk is unknown as P_f is unknown. Risk assessment is introduced to estimate the risk. We refer to P_f^* as the estimate (estimator) of P_f. By proper analysis, using data and modelling, the ambition is to determine the true value of P_f and hence the risk. This means that the error term $P_f^* - P_f$ must be sufficiently small so that it can be ignored. Statistical theory is the tool to be used to demonstrate the precision of the estimates. Confidence intervals are used as a measure of the uncertainty of the estimators relative to true value of the parameters. In this setting probabilities only exist as relative frequencies.

5.2 Case 1: Statistical inference of accident data

We refer to Section 4.1. Based on the statistics available on injuries, we will in this section perform a quantitative risk assessment with the aim of expressing injury risk for the coming year(s).

To define risk in this setting we introduce two indices,

$$F(x) = P_f(X \le x)$$

$$\lambda = E_f[X],$$

where X is the number of serious injuries during year 12. Hence F(x) is the relative frequency-interpreted probability that the number of injuries in year 12 does not exceed x, and λ is the expected number of serious injuries in the same year when interpreting the expectation as the average in an infinite population of similar years.

To be able to make meaningful comparisons for different time periods and activities we have to normalise the distribution and parameter with respect to manhours. Let c denote the number of million manhours in year 12. This number is assumed known. Let Y denote the number of serious injuries per million manhours, i.e.

$$Y = X/c.$$

Then we can define adjusted indices:

$$G(y) = P_f(Y \le y) = P_f(X/c \le y) = P_f(X \le yc) = F(yc)$$

and

$$\mu = E_f[Y] = \lambda/c.$$

Hence $G(y)$ is the probability distribution of the number of serious injuries per million manhours and μ is the expected number of such injuries.

This set-up is based on the assumption that there exist underlying true values of $F(x)$, $G(y)$, λ and μ. The aim of the risk assessment is to accurately estimate these distributions and parameters. To this end we need to have a clear understanding of the meaning of the true values. It is meaningless to try to accurately estimate these values if we cannot define them in a precise way. We focus on $F(x)$ and λ.

The true value of λ is the average number of serious injuries and $F(x)$ is the distribution of years with not more than x serious injuries as defined in the following model world:

Consider year 12. We construct in our mind an infinite population of "similar" years. For these years some aspects are allowed to vary whereas others must be viewed as fixed. We may specify the installations, the equipment type used, the operational procedures, the type of personnel positions, the type of training programmes, the organisational philosophy, the influence of exogenous factors, etc., but something must be different, because otherwise we would get exactly the same output result for each year: a fixed number of serious injuries. There must be some variation on a micro level to produce the variation from one year to the other. So we should allow for variations in the equipment quality, human behaviour, etc. But the question is, to what extent should we allow for such variation? For example, in human behaviour, do we specify the safety culture or the standard of the private lives of the personnel, or are these factors to be regarded as factors creating the variations from one year to another? If we consider the years 1–11 to be similar to year 12, this could indicate the type of population we consider. For these years there has obviously been some variation with respect to equipment quality, culture, management, influence of exogenous factors, etc. However, this population comprises just 12 years. To define the parameters we need to extend this population to a very large population (in theory, an infinite population). If some of these years are characterised by a somewhat different management climate than the rest of the observation period, we have to clarify to what extent the whole population should have the same balance with respect to years with a climate as we have in the period 1–12. Obviously if we allow for some time trend this balance is not maintained. We see that this discussion leads to different models depending on the understanding of the population of years to be considered. Below we will consider two types of models, one where we assume no trend and one where we assume a linear trend.

5.2.1 No trend

Let $X_1, X_2, \ldots, X_{12}$ be random variables representing the number of serious injuries for the years 1, 2, …, 12, respectively. Hence $X = X_{12}$. The distribution of X_i is denoted $F_i(x)$. As a model of $F_i(x)$ we use the Poisson distribution with parameter λ_i. We assume that $\lambda_i = \mu c_i$, where c_i is the number of manhours (in millions) in year i and μ is the expected number of serious injuries per million manhours.

A Poisson distribution is often used for describing the number of events occurring during a specified period of time. The motivation is as follows: If Z has a binomial distribution with parameters n and p, with n large and p small, the binomial distribution can be accurately approximated by the Poisson distribution with mean np; cf. Appendix A.1.3. Consider the number of events Z in a time interval [0,t], and let us divide the interval into a number of small subintervals. Then we may ignore the probability of two or more events occurring in each subinterval, and the total number of events in [0,t] can be written as a sum of "successes" in a number of Bernoulli trials (in each trial the outcome is either success or failure, the probability of success is p in all trials and the trials are independent). It follows that Z has a binomial distribution with large n and small p, and can consequently be approximated by a Poisson distribution.

We assume that the random variables $X_1, X_2, \ldots, X_{12}$ are independent and then it follows that the total number of serious injuries during the first 11 years is a Poisson random variable with parameter

$$\Sigma \lambda_i = \mu \Sigma c_i, \qquad (5.1)$$

where the sum is over i = 1, 2, …, 11. This follows by extending the above motivating arguments for the Poisson distribution by considering a period of time equal to all the years 1–11. From formula (5.1) we are led to an estimator μ^* of μ by replacing the total expected number of serious injuries in this period of time by the total number of serious injuries, which we denote by X_{1-11}. Hence

$$\mu^* = X_{1-11}/\Sigma c_i. \qquad (5.2)$$

We see that the expected value of μ^* equals μ, as

$$E\mu^* = EX_{1-11}/\Sigma c_i = \Sigma \mu c_i/c_i = \mu.$$

Figure 4.1 shows the total number of serious injuries for the period of 11 years. The underlying data are shown in Table 5.1

Table 5.1 *The underlying data of Figure 4.1.*

Year	1	2	3	4	5	6	7	8	9	10	11
Manhours c_i (in millions)	30.8	29.4	31.1	32.1	31.8	32.2	32.3	33.6	35.4	37.8	39.7
Number of serious injuries X_i	40	51	70	70	52	45	30	42	38	36	34
Normalised number of serious injuries $Y_i = X_i/c_i$	1.30	1.74	2.25	2.18	1.64	1.40	0.93	1.25	1.07	0.95	0.86

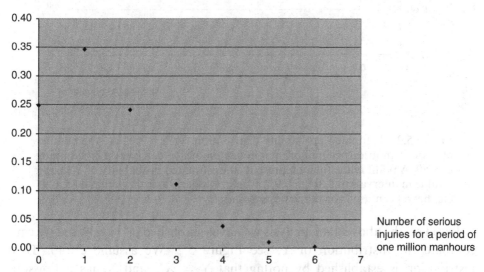

Figure 5.1 Estimated probability distribution for the number of serious injuries for a period of one million manhours, i.e. an estimate of $P_f(X=x)$ if $c=1$ using the Poisson distribution with parameter 1.39.

Using formula (5.2) this gives

$$\mu^* = X_{1-11}/\Sigma c_i = 508/366 = 1.39.$$

Based on this estimate we can compute an estimate of the probability distribution of a random variable representing the number of serious injuries for a period of one million manhours in the coming years; see Figure 5.1.

We see from the figure that the probability that the number of serious injuries is 0 is estimated to be about 25 per cent. Furthermore, the probability that the number of serious injuries is four or higher is estimated to be about 5 per cent.

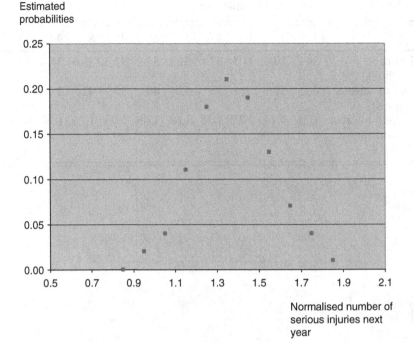

Figure 5.2   Estimated probability distribution for $Y = X/c$, i.e. the number of serious injuries per million manhours the next year (year 12) assuming $c = 40$. A point in the figure represents the estimated probability of Y taking a value in intervals $(0.8, 0.9]$, $(0.9, 1.0]$, etc. On the x-axis the values 0.85, 0.95 etc. have been used for the various points.

Based on the above data analysis we have established $\mu^* = 1.39$ and we can compute the distribution of Y; see Figure 5.2. We assume $c = 40$. This distribution is established by noting that $Y = X/c$ and X has a Poisson distribution with parameter $c\mu$, and estimated $c\mu^* = 40 \times 1.39 = 55.6$.

As seen from Figure 5.2, the distribution of Y is approximately a normal distribution, with mean 1.39 and variance $1.39/40 = 0.035$ as $\text{Var } Y = \text{Var}[X/c] = \text{Var}[X]/c^2 = c\mu/c^2 = \mu/c$.

Note that Y is the average of 40 variables each having a distribution as in Figure 5.1. Hence the approximately normal distribution is expected due to the central limit theorem (refer to Appendix A.1.3).

Next we need to address the accuracies of these estimators compared to the true underlying parameters.

Firstly we compute the variance of the estimator μ^* given by (5.2). We obtain

$$\text{Var}[\mu^*] = \text{Var}[X_{1-11}/\Sigma c_i] = (1/\Sigma c_i)^2 \text{Var}[X_{1-11}]$$
$$= (1/\Sigma c_i)^2 [\Sigma \mu c_i] = \mu/\Sigma c_i = \mu/366,$$

noting that X_{1-11} is a Poisson variable with parameter $\Sigma\mu c_i$.

By the so-called Chebyshev's inequality (Ross, 1993), we have

$$P(|\mu^* - \mu| \geq d) \leq \text{Var}[\mu^*]/d^2 = (\mu/366)/d^2 = \mu/(366d^2).$$

Thus the probability that the estimation error $|\mu^* - \mu|$ is greater than a (small) number d, say 0.30, is less than $\mu/(366d^2) = \mu/33$. The parameter μ is unknown, but the data strongly indicate that $\mu \leq 3$, and hence

$$P(|\mu^* - \mu| \geq 0.3) \leq 3/33 = 0.09,$$

i.e. the probability that the estimation error exceeds 0.3 is less than 10 per cent. In other words, we are confident that the estimator we use in this case produces estimates within a 0.3 error relative to the true underlying serious injury rate.

The accuracy can, however, be improved by using confidence intervals as explained in the following.

The aim is to compute a 90 per cent confidence interval for the unknown parameter μ. For this purpose we use the approximation to the normal distribution. We know that μ^* has an approximate normal distribution with mean μ and variance $\mu/366$ (standard deviation $\mu^{1/2}/19.1$) Hence

$$P(-1.65 \leq 19.1(\mu^* - \mu)/\mu^{1/2} \leq 1.65) \approx 0.90,$$

which gives an approximate 90 per cent confidence interval for μ equal to (1.29, 1.49).

Hence, assuming no trend we have estimated a serious injury rate per million manhours to be 1.39 and calculated an approximate 90 per cent confidence interval for this rate equal to (1.29, 1.49).

Next we would like to give a similar risk description for the fixed installations and the mobile units separately.

The resulting data are summarised in Table 5.2.

The data indicate that the risk level expressed by the serious injury rate is much lower for production installations than for mobile units. This can be confirmed by statistical hypothesis testing as shown in the following.

Let μ_M and μ_F be the normalised serious injury rates for the mobile units and fixed installations, respectively, defined analogously to μ. Furthermore let $(\mu_M)^*$ and $(\mu_F)^*$ be the estimators of μ_M and μ_F defined in line with formula (5.2). We see from Table 5.2 that $(\mu_M)^* = 1.93$ and $(\mu_F)^* = 1.23$.

We will test the hypothesis that the risk level for mobile units is higher than for fixed installations, based on the risk parameters μ_M and μ_F. The null hypothesis is that the parameters are equal, i.e.

$$H_0 : \mu_M = \mu_F.$$

Table 5.2 *Serious injury data for fixed installations and mobile units, with estimates and confidence intervals for* μ.

	Fixed Installations	Mobile units	All units
Number of serious injuries $\Sigma\ X_i$	359	159	508
Manhours $\Sigma\ c_i$ (in millions)	283.6	82.4	366.0
Normalised number of serious injuries μ^*	1.23	1.93	1.39
Var[μ^*]	$\mu/284$	$\mu/82$	$\mu/366$
Approx. confidence interval for μ, the serious injury rate per million manhours	(1.12, 1.34)	(1.68, 2.18)	(1.29, 1.49)

The alternative hypothesis is that the serious injury rate for mobile units is higher than for fixed installations, i.e.

$$H_1 : \mu_M > \mu_F.$$

We assert that H_1 is true if $(\mu_M)^* - (\mu_F)^*$ is "large". From Table 5.2 we note that $(\mu_M)^* - (\mu_F)^* = 0.6$, and we can easily compute the significance probability by using that $(\mu_M)^* - (\mu_F)^*$ has an approximate normal distribution. The variance of $(\mu_M)^* - (\mu_F)^*$ is given by the sum of the variances of the two estimators (assuming independence), i.e. $(\mu_M/82) + (\mu_F/284)$. Replacing the μ parameters by their estimators we obtain an estimator of the variance, and we can compute the significance probability

$$P((\mu_M)^* - (\mu_F)^* \geq 0.6|H_0) =$$
$$P([(\mu_M)^* - (\mu_F)^*]/SD^*] \geq 0.6/0.17 = 3.5|H_0),$$

where SD* is the estimated standard deviation of $(\mu_M)^* - (\mu_F)^*$, which in this case is equal to the square root of $(1.93/82) + (1.23/284)$.

From statistical tables we find the probability that a standard normal distributed random variable exceeds 3.5, namely 0.0002. Hence the significance probability is not larger than 0.02 per cent and the data provide strong evidence for concluding that H_1 is true.

5.2.2 Linear trend. Regression analysis

In this section we will use a linear trend analysis to predict the serious injury rate for the coming years. Firstly we perform the analysis for fixed installations. Table 5.3 shows the data corresponding to Figure 4.2.

Table 5.3 *The underlying data of Figure 4.2 (fixed installations).*

Year	1	2	3	4	5	6	7	8	9	10	11
Manhours c_i (in millions)	23.7	22.0	23.6	23.8	24.7	26.9	26.6	26.2	27.9	29.0	29.1
Number of serious injuries X_i	28	35	49	45	38	36	21	29	24	25	19
Normalised number of serious injuries $Y_i = X_i/c_i$	1.18	1.59	2.07	1.89	1.54	1.34	0.79	1.1	0.86	0.86	0.65

As in the previous section we let X_1, X_2, ... , X_{12} be random variables representing the number of serious injuries for the years 1, 2, ..., 12, respectively. The distribution of X_i is denoted $F_i(x)$. As a model of $F_i(x)$ we use the Poisson distribution with parameter λ_i. We assume that $\lambda_i = \mu_i c_i$, where c_i is the number of manhours (in millions) in year i and μ_i is the expected number of serious injuries per million manhours in year i. The normalised number of serious injuries in year i, X_i/c_i, is denoted Y_i. We assume that μ_i has a linear form

$$\mu_i = \alpha + \beta i,$$

where α and β are unknown parameters. To estimate these parameters we use standard regression analysis (see Appendix A.2.1). Using the method of least squares, i.e. we identify the values that minimise the sum of squared errors in the sample, we obtain the following estimators for α and β:

$$\alpha^* = \overline{Y} - \beta^*,$$
$$\beta^* = \Sigma_i(Y_i - \overline{Y})(U_i - \overline{U})/\Sigma_i(U_i - \overline{U})^2,$$

where $U_i = i$, and $\overline{U}$ and $\overline{Y}$ are the means of the U_i and Y_i, respectively, $i = 1$, 2, ..., n (n = 11). To predict Y we use the line

$$Y = \alpha^* + \beta^* i.$$

Figure 5.3 shows the serious injury rate data for fixed installations and the estimated regression line. Based on the line, an estimated expected serious injury rate for year 12 is 0.63. This number may also be seen as a prediction of the actual rate at year 12. We remember the rate based on the analysis in the previous section, which produced an estimate of 1.23 serious injuries per million manhours. Thus the trend analysis gives a much lower estimate, which is evident from the estimated linear curve in Figure 5.3.

Confidence intervals can be established for the parameters α, β and μ_i. The slope of the line, β, is of special interest as it is a measure of the trend

Accurate risk estimation

Serious injury rate

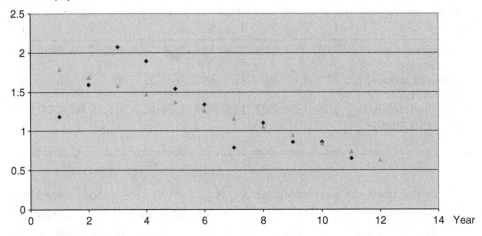

Figure 5.3 Serious injury rate for fixed installation and the estimated regression line.

of the data. If $\beta = 0$ there is no trend and often the analysis is concerned about the extent to which the data prove that there is a trend present. Could the observed decrease in the slope just be a result of "randomness"? As an approximation we assume that Y_i has a normal distribution. This gives a good approximation as $Y_i = X_i/c_i$ and X_i is close to a normal distribution when the parameter of the Poisson distribution is large as in this case. A 90 per cent confidence interval for β is then given by (Appendix A.2.1).

$$\beta^* + / - t_{n-2} \cdot S_\beta,$$

where t_{n-2} is the 95 per cent quantile of the Student (t) distribution with $n-2$ degrees of freedom, and S_β is an estimator of the standard deviation of β^* given by

$$(S_\beta)^2 = \left[\Sigma_i(Y_i - \alpha^* - i\beta^*)^2/(n-2)\right]/\Sigma_i(X_i - \overline{X})^2.$$

When computed, we obtain the following interval: $-0.10 +/- 1.81 \cdot 0.032 = (-0.16, -0.04)$. Hence, the data provide strong evidence that the true β is negative.

Similarly, a confidence interval for $\mu_i = \alpha + i\beta$ can be formulated. It takes the following form (Berenson *et al.*, 1988, p. 399):

$$\alpha^* + i\beta^* + / - t_{n-2} \cdot \left[\Sigma_j(Y_j - \alpha^* - j\beta^*)^2/(n-2)\right]^{1/2}$$
$$\{(1/n) + (X_i - \overline{X})/\Sigma_i(X_i - \overline{X})^2\}^{1/2}.$$

Numerically this gives for the coming year $i = 12$:

$$0.63 + / - 1.81 \times 0.13 = (0.39, 0.87).$$

We can conclude that the data provide a rather narrow uncertainty band for the expected serious injury rate for year 12.

One may also attempt to develop a prediction interval for the actual serious injury rate next year, as in Berenson *et al.* (1988, p. 401), but this interval would be a prediction interval *estimate*, as the probabilities generating the interval are unknown and must be estimated.

5.3 Case 2: QRA of the LNG plant

We refer to Section 4.2 where we introduced Case 2, addressing the risk related to a new Liquefied Natural Gas (LNG) plant located in an urban area. Our focus is the third-party risk, i.e. risk for the people living, working or staying outside the Lyse LNG base load plant. Two risk indices are defined:

- Individual risk (IR), defined as the probability that a specific person (arbitrarily chosen) shall be killed due to the activity during a period of one year.
- The f-n curve, expressing the frequency (i.e. the expected number) f of accidents that leads to minimum n number of fatalities, which can also be interpreted as the probability of an accident with at least n fatalities.

The probabilities are frequentist probabilities and are interpreted as the fraction of times the event studied occurs if the situation could be repeated infinitely under similar conditions. The f-n curve is interpreted as the average number of accidents leading to minimum n fatalities, when considering an infinite number of similar situations. The average is approximately equal to the fraction of situations with at least n fatalities.

In the risk analysis, estimates of these indices are computed, and these estimates are compared to the risk acceptance (tolerability) limits defined; see Section 4.2.

Let

$p = $ IR for a specific person in the group having the highest risk

Furthermore let $Y(n)$ denote the number of accidents with at least n fatalities during a specific period of time (for example one year). Then the f-n curve is defined by $E_f[Y(n)]$ and $E_f[Y(n)] \approx P_f(Y(n) \geq 1)$. If Z is the actual number of fatalities in the same period and we can exclude the probability of two or more accidents, we have

$$P_f(Y(n) \geq 1) \approx P_f(Z \geq n).$$

Let

$$G(n) = E_f[Y(n)].$$

The aim of the risk assessment is to accurately estimate p and G(n).

5.3.1 Outline of the main features of the risk assessment

The risk assessment covers four main steps (see Figure 1.2):

1. Hazard identification (e.g. a gas leak)
2. Cause analysis (including a hazard frequency analysis)
3. Consequence analysis
4. Risk picture.

In the following we will sketch the assessment for these four steps, first the hazard identification.

Hazard identification

The first step of the execution part of a risk assessment is the identification of initiating events. If the focus is on hazards (threats), as in our case, then we are talking about a hazard identification (threat identification). It is often said that "what you have not identified, you cannot deal with". It is difficult to avoid or to protect oneself against hazards and threats that one has not identified. For this reason, this part of the analysis is a critical task of the assessment. However, care has to be shown to avoid this task becoming routine. When one performs similar types of analyses, it is common to copy the list of hazards and threats from the previous assessments. By doing this, one may overlook special aspects and features of the system being considered. It is therefore important that the identification of initiating events be carried out in a structured and systematic manner, and that it involves persons having the necessary competence (Aven, 2008a).

The development of the list of initiating events is based on different types of input, including similar types of analyses as mentioned above, general experience, databases, inspections and assumptions. Special techniques are often used, for example Hazard and Operability studies (HAZOP) (Aven, 2008a). A common feature in all the methods is that they are based on a type of structured brainstorming in which one uses checklists, guide words, etc., adapted to the problem situation being studied.

For the LNG case study, the analysts first identified the major inventories of flammable and explosive materials in the LNG plant units, together with the major lines connecting the inventories. From this a set of hydrocarbon leaks, including pipeline ruptures, of different sizes at various locations were identified. These have a potential for fire or explosion. "Non-hydrocarbon leakage hazards", such as occupational accidents, were not addressed in the QRA. These hazards were analysed through other techniques, such as the HAZOP study.

In the case study a crude qualitative risk analysis was carried out, identifying the hazards, the causes, consequences and safeguards. As an example consider the hazard "LNG leakage from process equipment or piping". Causes identified were corrosion, erosion and gasket failure, and possible consequences were flashing and evaporation of LNG, atmospheric dispersion of gas, potential explosion, flash fire, jet fire and pool fire. The safeguards listed were regular maintenance, fire and gas detection, emergency shutdown system, emergency blowdown system, and active/passive fire protection.

Cause analysis (including a hazard frequency analysis)

In the cause analysis, we study what is needed for the initiating events to occur. What are the causal factors? Several techniques exist for this purpose, from brainstorming sessions to the use of fault tree analyses and Bayesian networks (Aven, 2008a). Experts on the systems and activities are usually necessary for carrying out the analysis. Having established an understanding of the causal factors, the frequencies or probabilities of the hazards (for example, the hydrocarbon leakages) can be estimated. Often the analyses have a main focus on this estimation. This was the case in the LNG case. Leakage causes were identified, such as corrosion and erosion, as mentioned above, but a detailed analysis of the causes was not performed.

For each initiating event (hydrocarbon leakage) a frequency was estimated. This was done by using the so-called parts count method. The main ideas of this method will be sketched in the following.

A potential leakage could come from several sources, e.g. valves, flanges, vessels, compressors, etc. As an example, consider leakages from valves. The first step will be to count the number of each type and size of the valves in the area of consideration, and to multiply this number with the leakage frequency for each valve type. This leakage frequency is estimated from a statistical database (data bank). Then we do the same for all other equipment categories, and finally we add them together to obtain the total leakage frequency in the area.

In the LNG case, four leak size categories were introduced:

Leak range category	(mm)	Representative size (mm)
small	1–10	3.3
medium	10–50	23
large	50–100	53
very large, full bore rupture	>100	124

Leaks with diameter below 1 mm and rate below 0.1 kg/s were not studied as they were not considered to contribute substantially to the overall risk.

The frequency of leakages for these categories was estimated based on a leak frequency database, the UK HSE Offshore Hydrocarbon Release Statistic 1992–2006 or HCRD (HSE, 2000). For example the following frequencies were estimated for a specific system (LNG refrigerant system):

Gas leak category	Estimated Frequency
small	2.3×10^{-3}
medium	3.1×10^{-4}
large	7.9×10^{-5}
very large, full bore rupture	1.0×10^{-4}
Total	2.5×10^{-3}

To reflect specific aspects of the LNG plant, which is considered a clean service and an onshore facility, leak frequencies for pipelines, vessels and the LNG Storage Tank are applied as given in the Purple Book (2008). Accordingly, failures of flanges are assumed to be included in the failure frequency of the pipeline. This reduces the leak frequency by deleting the flanges on pipelines and vessels (tanks). For example the frequency estimate for process pipe (1m) of 6×10^{-5} is changed to 5×10^{-6}.

Consequence analysis

In the consequence analysis we study the effects the initiating events A may have on human beings, the environment and financial assets (or something else that humans value). Scenarios are developed showing how the initiating events could lead to specific consequences, for example with respect to loss of lives. The typical analysis, as in our case, is to first analyse the physical consequences of the initiating events, for example the flow from liquids and gases. To determine the physical effect of an event, for example related to gas dispersion, explosion pressure, etc., special techniques and computer codes are normally used.

The immediate effects of an event can be determined in a coarse risk analysis as was mentioned above under the heading "hazard identification", but other techniques are required to develop the scenarios. In the actual case, event trees were used; see Figure 4.7 for an example. Event tree analysis is the most common method for consequence analysis in risk assessments, but other methods such as Bayesian belief networks are also used.

The number of stages in an identified scenario depends on the complexity of the safety and control-systems (barriers). In a process plant with a high level of monitoring and control/alarm systems, the number of stages in the series of events can be relatively high, before the effects "leave the system".

In the case study a software for the physical analysis was used to assess the consequences of the hazards. The following categories were used:

- Dispersing of Hydrocarbon Vapour Cloud
- Explosion
- Fireball
- BLEVE
- Flash Fire
- Jet Fire
- Pool Fire.

The actual outcome depends on source conditions like the type of fluid, temperature, pressure etc. and release phenomena. To illustrate some of the features and assumptions made in the consequences analysis in the LNG case study, we give below a summary of some key aspects of the analysis carried out (Lyse, 2008).

For modelling purposes, the releases were categorised as either instantaneous or continuous.

If a catastrophic failure of the shell of a vessel occurs the contents would be released very quickly (instantaneously). This type of failure has been modelled as a hemispherical cloud centred on the release location.

Releases from pipelines, flanges, pumps, etc. are modelled as liquid, gas, or two-phase releases. Where an inventory comprises significant liquid and gas sections, e.g. in a vessel, then both are modelled and analysed. The representative release height for all cases is taken as 1 m, except for the LNG Tank, where 30 m is applied, since the leak sources (flanges) by the LNG Tank are likely to be on the tank top.

Release rates are assumed to be constant throughout the release duration time and calculated with isolation (ESD System), and with blowdown.

According to specific standards the isolated sections shall be depressurised to 50 per cent of design pressure in 15 minutes or to 7 barg in 30 minutes. Based on this, a reaction time from the first fire & gas alarm until the operator initiates the ESD and blowdown system is assumed to be 600 seconds. An average blowdown time of 900 seconds is used in the calculation.

For each leak category an initial release rate (assuming continuous release) can be calculated by standard release formulae (see e.g. Aven, 1992, p. 212). For a specific hole size (area of release plane), the initial release rate is expressed as a function of pressure and temperature.

When a vapour cloud is generated, either instantaneously or continuously, there may be a substantial degree of mixing of air with the released material. To allow for destruction of momentum due to impingement of releases or upwind and downwind releases, 50 per cent of releases were modelled as free-field horizontal releases and 50 per cent were modelled as "impinged" releases. The dimensions of impinged releases were determined assuming that the clouds were cylindrical in shape, but with the same volume as a horizontal release.

In the event of a release from containment which is not ignited immediately, a hydrocarbon vapour/air mixture is formed. The concentration of hydrocarbon in the cloud, as progressive dilution with air takes place, is estimated using a dispersion model. The direction and extent of drift of the cloud is influenced by the prevailing weather conditions. The cloud remains capable of ignition providing the concentration is above the lower flammable limit (LFL). On ignition, a flame front passes at slow speed throughout the flammable cloud and a flame stabilises near to the point of release as either a jet or pool fire. Jet fires are usually the consequence of a momentum-dominated release resulting from an immediately ignited release or from a flash fire that burns back to the point of release.

Release of flammable fluid may have many outcomes, depending on the timing and type of ignition. For example, a release may ignite immediately at the point of release, or it may ignite after the cloud has been dispersing for two minutes, or after the cloud has been dispersing for five minutes, or it may not ignite at all. If it ignites, it may give either explosion effects or different types of fire effects depending on the type of release (e.g. jet fire, fireball, pool fire or flash fire).

The different outcomes are presented in the form of event trees, as in Figure 4.7. The immediate ignition probability is directly specified. A default value of 0.3 is used. The delayed ignition probability for any failure case is based on the defined ignition sources on site, with a unique value for each release case and release direction. The calculation is based on the strength, location and presence factor of all ignition sources specified, and the size and duration of the dispersing flammable vapour cloud.

To study the effect of a release, weather data, in particular the wind rose data, provide useful information. These data show the fraction of time that the wind has a specific direction and speed.

To calculate the number of fatalities a population distribution is determined. For the various areas the number of people exposed is specified, for days and nights. For example at the ferry terminal, about 1500 people are indicated for the daytime and 0 during nights.

For each scenario a fraction of fatalities are determined. For example for heavy explosion scenarios a fraction of 1 was used whereas for many types of fires a fraction of 0.2 was used. For some detailed calculations, see Section 6.3.

Risk picture

The calculations produce estimates p^* and $G^*(n)$ of

$p = IR$ for a specific person in the group having the highest risk

and the f-n curve $G(n) = E_f[Y(n)]$, where $Y(n)$ denotes the number of accidents with at least n fatalities during a specific period of time (for example one year). Figure 4.6 shows the f-n curve estimate $G^*(n)$. We see that for small n values the estimate $G^*(n)$ lies in the ALARP region, whereas for large values of n the risk is acceptable.

The aim of the risk assessment is to accurately estimate p and $G(n)$.

The estimate p^* is equal to 5×10^{-7}, i.e. the estimated risk is in the ALARP region, as the following limits were defined (refer to Figure 4.5):

$$IR > 1 \times 10^{-5} \quad \text{Not acceptable (intolerable)}$$
$$1 \times 10^{-5} < IR < 1 \times 10^{-7} \quad \text{ALARP}$$
$$IR < 1 \times 10^{-7} \quad \text{Acceptable risk}$$

Hence, the risk is not considered unacceptable or intolerable. In applications the risk estimates typically fall in the ALARP region, close to acceptable.

Several risk-reducing measures were considered, for example welded pipes in feed gas and LNG services. The measures reduced the estimated risk to a level very close to the acceptable region. The assessments were carried out in parallel with sensitivity analyses of some assumptions, including the personnel distribution and the fraction of fatalities in different types of scenarios.

5.4 Case 3: Design of a safety system

We refer to Section 4.3 where we introduced Case 3, the design of a safety system. The reliability and risk indices introduced are:

λ: the failure rate of the unit, the number of failures per unit of time when considering a sequence of a very large (in theory infinite) number of similar units replaced at failures.

MFDT (mean fractional dead time) for the two systems, interpreted as the fraction of time the system is not functioning in the long run.

q = IR for a specific person in the group having the highest risk, for the two systems.

The aim of the assessment is to accurately estimate these quantities.

For the unit we suppose that the following failure data are available:

Source a: 10 failures with a total exposure time of about 500
Source b: 5 failures with a total exposure time of about 25.

The data source b is considered to have the most relevant data for the systems analysed, but there are different views on this among the analysts performing the analysis. Using the data from source b, we obtain an estimate λ^* of λ equal to $5/25 = 0.2$, as was indicated in Section 4.3. Using source a, we obtain an estimate of λ equal to $10/500 = 0.02$, a factor 10 times lower than λ^*.

Using the standard formulae for MFDT, $\lambda\tau/2$ (alternative I) and $(\lambda\tau)^2/3$ (alternative II) and that the time between tests is $\tau = 1$, we obtain MFDT estimates equal to 0.1 (10%) for alternative I and 0.013 (1.3%) for alternative II when λ is replaced by the estimate $\lambda^* = 0.2$. If the estimate 0.02 is used for λ, the corresponding estimates of MFDT are 0.01 and 0.00013.

The safety system influences the risk, and a risk assessment similar to the one described in Section 5.3 is conducted and estimates q^* of q are obtained. Using $\lambda^* = 0.2$ the q estimates are 4×10^{-4} and 2×10^{-4}, respectively for alternative I and alternative II. The difference is moderately large, a factor of two, as the failure of the safety system could affect the escalation of the accident. If the estimate of 0.02 of λ is used, the corresponding estimates of q are 1×10^{-4} and 2×10^{-5}, i.e. the risk index for alternative II is a factor of 5 lower than the one for alternative I.

The decision-maker compares this difference with the extra costs of implementing alternative II.

5.5 Discussion

In this section we will discuss the scientific quality of the assessment using the reliability and validity requirements (refer to Section 3.3). The three examples above will be used to illustrate the analysis. Of the validity criteria, only V1 and V4 are applicable.

5.5.1 Validity criterion V1: Accurate risk estimations

Firstly let us look into the validity criterion V1: the degree to which the produced risk numbers are accurate compared to the underlying true risk. This is obviously an appropriate criterion in this case as the aim of the analysis is to accurately estimate risk. Relative frequency-based probabilities $p = P_f$ or related parameters are assumed to exist, generating a "true" risk (A, C, P_f), as defined in Section 5.1.

If we have available a substantial amount of data and these are considered relevant for the estimation of p, i.e. the observations are considered similar to those of the population generating p, statistical theory shows that the estimation error becomes negligible. Hence the results are valid according to the criterion V1.

It is not possible to define precisely what we mean by terms such as "accurate estimation" and "negligible estimation error" without being explicit about the context. However, often we may indicate an order of magnitude. For example, if we estimate a p equal to 0.10 and the upper and lower confidence bounds are 0.11 and 0.09 respectively, the estimation error would be considered negligible for most applications.

Consider Case 1: Statistical inference of accident data. In this case the confidence intervals are quite narrow; see for instance Table 5.2. For example the approximate 90 per cent confidence interval of μ, the serious injury rate per million manhours, equals (1.12, 1.34) and (1.68, 2.18) for fixed installations and mobile units, respectively. There is a considerable amount of data to conclude on the true risk numbers.

If the time period had been smaller, the intervals would of course have been wider. For example, if we had only one year as the basis for our analysis, the intervals for all units would be approximate estimates $+/- 0.35$. Then it would for example be more difficult to show significant differences between two rates.

Also in the case of linear regression, the analysis produces quite narrow uncertainty bands. For example, the 90 per cent confidence interval for $\mu_i = \alpha + i\beta$ for year 12 is equal to (0.39, 0.87).

However, in many practical risk analysis settings data are often scarce. If the statistical analysis is based on few data, the estimation accuracy would be poor as the confidence intervals would be wide. Thus accurate estimation, and a high validity according to V1, cannot be obtained.

Consider Case 2: QRA of the LNG plant. In this case estimates of the true underlying risk have been produced, but no attempt has been made to describe the uncertainties in the estimates. Intuitively, it seems clear that the estimates would be subject to large uncertainties, and the validity criterion V1 cannot be met.

It would be impossible to use confidence intervals to reveal these uncertainties as the estimates are not produced by data of the form X_1, X_2, ... as in case 1. For some parameters we may have used data of this form, but for most parameters the estimates are based on the analysts' (experts') best judgements given their knowledge at the time of the assessment. In the next chapter we will study the probability of frequency approach (see Section 2.7) which allows for uncertainty descriptions of the risk parameters, using knowledge-based (subjective) probabilities. The aim then is to describe the uncertainties, not to accurately estimate the risk, but we do not need detailed analysis to conclude that the criterion V1 cannot be satisfied if the aim is accurate risk estimation in a case like this. There are too little relevant data available, and the models used introduce an additional uncertainty element.

Case 3: Design of a safety system, is similar to case 2, but it would be easier to express uncertainties about the risk parameters, λ, MFDT and q. Based on the data from source a and b, we can produce confidence intervals for the failure rate λ, and this would in turn lead to confidence intervals for MFDT and q, reflecting the failure data of the unit. Suppose that the data from source b, five failures with an exposure time of about 25, is generated by observing five failures in a time period 25 with only one unit of exposure at any time. Then the number of failures has a Poisson distribution with parameter 25λ, and a 90 per cent confidence interval is given by (see Appendix A.2.1):

$$(z_1/50, z_2/50) = (3.94/50, 21.03/50) = (0.08, 0.42),$$

where z_1 equals the 0.05 quantile in the chi-square distribution with $2 \times 5 = 10$ degrees of freedom, and z_2 equals the 0.95 quantile in the chi-square distribution with $2 \times 6 = 12$ degrees of freedom. The values of z are found from statistical tables for the chi-square distribution. This gives the following 90 per cent confidence intervals for MFDT for the two alternatives:

Alternative I: $(0.08/2, 0.42/2) = (0.04, 0.21)$
Alternative II: $(0.08^2/3, 0.42^2/3) = (0.0021, 0.059)$.

We have used the fact that MFDT is given by $\lambda\tau/2$ and $(\lambda\tau)^2/3$ for these alternatives and the time between tests is $\tau = 1$. The MFDT estimate computed above was 0.1 and 0.013 respectively. We may also compute approximate confidence intervals for q = IR, using the calculations in the risk assessments. The intervals would then cover the estimates 4×10^{-4} and 2×10^{-4}, for the two alternatives. As the risk calculations are not presented here, we just report the figures

Alternative I: $(3 \times 10^{-4}, 5 \times 10^{-4})$
Alternative II: $(1 \times 10^{-4}, 3 \times 10^{-4})$.

We see that the intervals are quite wide for λ and MFDT. For $q = IR$ the intervals are however rather narrow. The reason is that the IR value is governed by many other factors than the reliability of the safety system, and the interval only relates to variation in the data for this system.

We may discuss to what extent the estimation is accurate for λ and MFDT, but it is clear that with more observations the confidence intervals become narrower. If we include all data, we ignore the problem of relevancy of the data (refer to discussion of criterion V4 below); the corresponding 90 per cent confidence interval for λ is

$$(z_1/1050, z_2/1050) = (18.5/1050, 46/1050) = (0.017, 0.04),$$

i.e. a much more narrow interval than for source b.

To increase the amount of data, to produce more narrow confidence intervals, we have extended the relevant population of observations to cover situations that to a varying degree are similar to the one being studied. This reduces the quality, i.e. the relevancy of the data, but this aspect would not be possible to describe by the statistical analysis. If the data are not considered relevant, the statistical analysis cannot be used to check the validity according to criterion V1.

The same type of problem arises in the case of modelling, although the amount of data is often larger on the detailed system level. However, in this case we should also take into account the inaccuracy (uncertainty) introduced by using the model, for example that $MFDT = (\lambda\tau)^2/3$ in alternative II in Case 3. This model is based on independence which may be a rough assumption. However, the statistical analysis is not able to describe this uncertainty. The analysis is conditional on the model used. In the assessment we use a specific model and this introduces a possible error in the risk estimation. Hence validity V1 is ensured only if the model inaccuracy is negligible.

We conclude that the risk assessment when founded on traditional statistical methods meets the validity requirement according to V1 only if a large amount of relevant data is available. In other cases, when such data are not available, the V1 criterion is in general not satisfied.

Next we look into criterion V4. The criteria V2 and V3 are not relevant for this case, as already mentioned.

5.5.2 Validity criterion V4: Addressing the right quantities

Next we address the criterion V4: the degree to which the analysis addresses the right quantities. Are μ, p, $G(n)$, λ, q etc. really the quantities of interest?

Our goal is to express the risk of an activity, but in this set-up we are concerned about the average performance of a thought-constructed population of similar situations. Are these quantities meaningful representations of the system or activity being studied?

Let us investigate in more detail a parameter in the first case, the serious injury rate μ. We interpret this rate as the average number of serious injuries per million manhours when considering an infinite number of similar situations to the one studied. This interpretation we discussed in Section 5.2 for the analogous rate at year 12, λ_{12}. We had problems in explaining the meaning of this parameter. It expresses in a way some average number of serious injuries, but what the average covers is difficult to define and communicate. The idea is that the parameter reflects some underlying property of the activity studied. It is acknowledged that the meaning of this property is not fully understood but conducting the assessment just as if this property exists has proved to give a useful tool for practical analysis. However, this type of pragmatism cannot be accepted if we require that our assessment should be built on a scientific basis. Then all quantities introduced should be properly defined and allow for meaningful interpretations.

But even if we accept the existence of these parameters one may question the focus on these quantities in the risk assessment. Why should we focus on the average performance, and not the performance of the specific activity analysed? Would it not be more interesting to address the observable quantities, for example the number of serious injuries in the coming year?

No, is the answer when adopting the traditional statistical paradigm which constitutes the basis for the analysis in this chapter. Observations are subject to randomness, and we should search for the underlying parameters describing the condition of the system or activity analysed. This underlying condition is what we look for. Risk is related to this condition, not to the values of the random variables. Without such parameters, the assessments cannot be performed.

The strength of this argumentation relies on the ability to define models with parameters that are meaningful for characterising the specific unit studied. Only if the parameters have meaningful interpretations can we make a judgement about their appropriateness for describing risk. For all the three cases studied above the interpretation was difficult. The easiest case was the failure rate of the units studied in Case 3. But also in this case it is hard to define precisely the relevant population of units. Hence, we conclude that the validity requirement V4 is not in general met.

5.5.3 *The reliability criteria R – the same results with repetition*

The amount of relevant data also affects the reliability criterion R, the extent to which the risk analysis yields the same results when repeating the analysis. In the case of a large amount of relevant data, the statistical analyses would show insignificant variations from analysis to analysis for all three interpretations R1–R3:

> The degree to which the risk analysis methods produce the same results at reruns of these methods (R1).
>
> The degree to which the risk analysis produces identical results when conducted by different analysis teams, but using the same methods and data (R2).
>
> The degree to which the risk analysis produces identical results when conducted by different analysis teams with the same analysis scope and objectives, but no restrictions on methods and data (R3).

The statistical methods for such applications are largely universal. In other cases, when such data are not available, the R criterion is in general not satisfied. It is, however, not difficult to identify examples where the criteria R (and R1–R3) are also met when rather few data exist. If we have success – failure observations, for example two successes out of ten – the analysis teams are led to a binomial probability model and this would give the same estimate and confidence interval. For other situations there may be no obvious analysis methods and different analysis teams are likely to produce different results. If repeating the analysis means making a new data sample, this may lead to large variations in the results in the case that the sample is small (relevant for R1 and R3). For example, in the binomial case with 10 trials the success estimate could show large differences from sample to sample.

In view of V1, the reliability criteria are all considered appropriate for the traditional statistical methods. The aim is to produce risk numbers close to the true risk and then we should require that the analysis results are not dependent on the analysis team and/or the methods and data used.

For Case 1 we may thus conclude that the reliability criteria are met. We could repeat the analysis and get more or less the same results. This could also happen for Case 3 as there is a rather common approach to the problem. The MFDT models are well established. However, there exist many adjustments of the basic models, and different analysts could obviously make different choices with respect to the data to be used in the analysis. Criterion R2 – that the risk assessment produces identical results when conducted by different analysis teams but using the same methods and data – should thus be met in these cases (Cases 1 and 3 up to MFDT level). We get the same conclusion for criterion R1, that the risk assessment methods produce the same results at reruns of these methods, provided the same data are used. However, criterion R3, that the risk assessment produces identical results when conducted by different analysis teams with the same analysis scope and objectives but no restrictions on methods and data, does not hold for Case 3, only for Case 1 where there are no discussions on which data to apply.

For Case 2 (and the IR part of Case 3) we can conclude that the criteria are not in general met. Criterion R3 is obvious. It is obvious that the risk analysis does not produce identical results when conducted by different analysis teams with the same analysis scope and objectives, but no restrictions on methods and data (R3). This has also been documented by several benchmarking exercises; see e.g. Lauridsen *et al.* (2001). We have to conclude negatively also for criterion R2. The risk analyses do not produce identical results when conducted by different analysis teams, but using the same methods and data (R2). The point is that the methods used need to be based on assumptions, and these assumptions would be different for different analysis teams.

For criterion R1, that risk analysis methods produce the same results at reruns of the methods used, the conclusion would depend on what is allowed to vary at reruns. If the assumptions are not fixed, the criterion is obviously not met. If, however, the assumptions are fixed, the variation may reflect only the reliability of the computation procedures for producing the risk numbers. We all know that computer codes and hand-calculations can fail, but proper quality assurance procedures should be able to solve this problem. For complex models, computational issues could easily lead to unreliability of the results if not taken seriously.

5.5.4 Summary of assessments and final remarks

Table 5.4 summarises the conclusions from the previous section.

The traditional statistical methods meet the reliability and validity criteria only if a large amount of relevant data is available. If such data do not exist, the criteria are not in general met. Only in specific cases would some of these criteria be met, as discussed above.

The validity criterion V4, that the analysis addresses the right quantities, is met to the degree that the model parameters are adequately characterising the units studied, see Section 5.5.2.

In this chapter we have restricted the analysis to relative frequency-interpreted probabilities and standard statistical analysis. It would also be

Table 5.4 *Summary of reliability and validity analysis. Y indicates that the criterion is met, N that it is not met, and Y/N that it is met under certain conditions. The boxes are empty in cases where the criterion is not relevant.*

Approach	Criterion								
	R	R1	R2	R3	V	V1	V2	V3	V4
Traditional statistical analysis, large amount of relevant data available	Y	Y	Y	Y	Y/N	Y			Y/N
Traditional statistical analysis in other cases	N	N	N	N	Y/N	N			Y/N

possible to use Bayesian methods and subjective (knowledge-based) probabilities to assess uncertainties about the presumed underlying risk and relative frequencies. However, such methods and perspectives are more commonly adopted when the objective of risk assessment is to describe uncertainties, and not to accurately estimate risk. We refer to the next chapter.

6

Risk assessment when the objective is uncertainty descriptions

Next we will study the scientific platform of risk assessments when the objective of these assessments is to describe uncertainties. We follow the same structure as in the previous chapter. We first summarise the framework introduced in Chapter 2 for assessing risk in such a setting, and clarify key concepts like probability and risk. We then conduct the assessments for the three cases, and from this basis we study the scientific quality of the risk assessments. Focus is again on the scientific requirements reliability and validity defined in Chapter 3. We distinguish between an (A,C,P_f)-based risk perspective (referred to as the probability of frequency approach) and an (A,C,U)-based risk perspective; in the former risk is defined through chances (which is the Bayesian term for frequentist probabilities, i.e. fractions of "successes" in the long run; refer to Chapter 2) and in the latter risk is defined through uncertainties.

6.1 Scientific basis

We consider an activity and distinguish between the following two ways of looking at risk:

I: Risk is defined through chances (frequentist probabilities)

Risk $= (A,C,P_f)$, where P_f is a chance (relative frequency-interpreted probability) or a related parameter such as the expected number of occurrences of the event A per unit of time, where expectation is with respect to the chance distribution (relative frequency-interpreted probability distribution).

Risk is described according to the probability of frequency approach as presented in Chapter 2, i.e.

$$\text{Risk description} = (A, C, P_f*, P(P_f), K),$$

where K is the background knowledge that the estimate P_f* and the subjective (knowledge-based) probability distribution P is based on.

II: Risk is defined through uncertainties

Risk = (A,C,U), where U is uncertainties about A and C (will A occur and what will the consequences C be?).

The corresponding risk description based on this definition is

$$\text{Risk description} = (A, C, U, P, K),$$

where P is a knowledge-based (subjective) probability expressing the uncertainties based on the background knowledge K. This description covers probability distributions of A and C, as well as predictions of A and C, for example a predictor $C*$ given by the expected value of C, unconditionally or conditional on the occurrence of A, i.e. $C* = EC$ or $C* = E[C|A]$. The U represents some type of uncertainty analysis that "extends beyond P", for example a special identification and assessment of uncertainty factors as was discussed in Section 2.9.

In the following we study the three cases 1–3 for these two perspectives on risk.

6.2 Case 1: Statistical inference of accident data

6.2.1 Risk is defined through chances (frequentist probabilities)

We refer to Sections 4.1 and 5.2. In Section 5.2 we used the statistics available on injuries to estimate unknown risk parameters, for example the serious injury rate μ. To express uncertainties, confidence intervals were used. The aim was to accurately assess the risk, i.e. the parameters of the risk model. Now we change focus. The objective of the risk assessment is to describe the uncertainties about the parameters, and the tool used for this purpose is knowledge-based (subjective) probabilities P. These probabilities are conditional on a background knowledge K.

The analysis then becomes a standard Bayesian analysis, which is based on the following steps:

1. Establish a probabilistic model.
2. Assign a prior distribution on the parameters of interest.
3. Use Bayes' theorem to establish the posterior distribution of the parameters.

Two different probabilistic models were introduced in Section 5.2. For both models it is assumed that the number of serious injuries follows a Poisson distribution, but for the first model the rate is independent of time, whereas for the second the rate is a linear function of time (linear regression model). Consider first the time-independent model. The aim is then to assess the uncertainties about the true underlying serious injury rate μ.

As a prior distribution on μ we may use a gamma distribution f with parameters α and β:

$$f(\mu \mid K) = \beta(\beta\mu)^{\alpha-1}e^{-\beta\mu}/\Gamma(\alpha) \ \mu > 0,$$

where K denotes the background knowledge that the assessment is based on. This knowledge might come from various sources, e.g. observations from previous years, similar data from other countries, expert judgements, etc. The gamma distribution is a mathematically convenient choice for prior distribution in this case as it a so-called conjugate distribution. A prior distribution is called conjugate if it leads to a posterior distribution in the same distribution class (in this case gamma) (Singpurwalla, 2006).

Given that the total number of serious injuries in the previous 11 years X_{1-11} equals x, the posterior density is given by

$$f(\mu \mid x, K) = d \ L(\mu)f(\mu \mid K),$$

where d is a normalising constant such that the density has integral 1 and $L(\mu)$ is the likelihood function (which equals the probability density $p(x|\mu)$ of X_{1-11} when μ is known, seen as a function of μ):

$$L(\mu) = p(x \mid \mu) = c^x(\mu^x/x!)e^{-\mu c},$$

and $c = \Sigma \ c_i$.

We find that

$$f(\mu \mid x, K) = d\mu^{x+\alpha-1}e^{-(\beta+c)\mu},$$

i.e. the posterior density is a gamma distribution now with parameters $\alpha + x$ and $\beta + \Sigma \ c_i$.

Suppose the analyst's (expert's) prior knowledge indicates that it is not likely that we experience more than three serious injuries per million exposed hours. The probability that the parameter μ exceeds three should be about 10 per cent. The expected value is set to 1.5. This gives $\alpha = 2$ and $\beta = 4/3$. Note that the expected value equals $\alpha/\beta = 1.5$. The distribution (density) is shown in Figure 6.1, together with the posterior distribution (density) when the number of serious injuries $x = 508$ and the manhours equals 366 (remember Table 5.2). The posterior distribution is an approximate normal

Densities

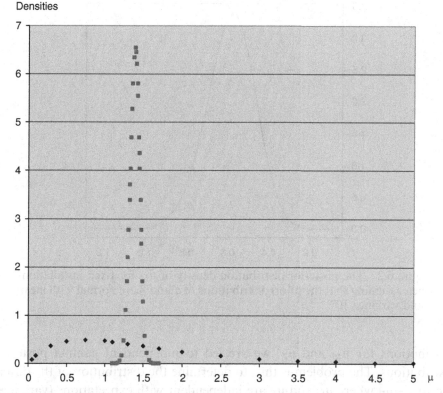

Figure 6.1   Prior (flat curve) and posterior (peaked curve) densities of the serious injury rate μ.

distribution, with the mass concentrated on (1.29, 1.49), which is a 95 per cent credibility interval for μ. The mean of this distribution is $(\alpha + x)/(\beta + \Sigma c_i) = 1.39$ and the variance equals $(\alpha + x)/(\beta + \Sigma c_i)^2 = 0.00378$ (standard deviation 0.061). Hence the analyst (expert) is 95 per cent sure that μ lies in this interval given the data observed for the 11 years. We see that the data observed dominate the posterior distribution. The posterior distribution is not very much dependent on the prior distribution in this case since the amount of data is so large. The credibility interval is approximately equal to the confidence interval calculated in Section 5.2. The meanings of these intervals are, however, different. We refer to the discussion in Section 6.5.

Also the credibility intervals for fixed installations and mobile units are approximately equal to the computed confidence intervals (see Table 5.2) for "reasonable" choices of the prior distributions.

The posterior analysis can also be used to determine the probability that the serious injury rate for mobile units (μ_M) is larger than for fixed installations (μ_F), i.e. $P(\mu_M > \mu_F \mid x, K)$. Using independent prior gamma

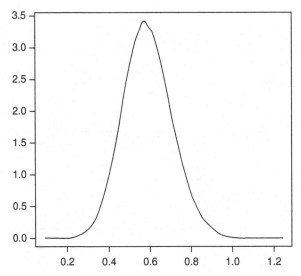

Figure 6.2   The posterior distribution density of μ_{12} for fixed installations when assuming that the priori distributions of α and β are normal with mean 0 and variance 10^6.

distributions for μ_M and μ_F, we are led to independent gamma posterior distributions. The problem is thus to determine the distribution of the difference $\mu_D - \mu_F$ where μ_D and μ_F are independent with expectations (variance) approximately equal to 1.93 and 1.23 (1.93/82 and 1.23/283), respectively. Using approximation to the normal distribution, we find that

$$P(\mu_M - \mu_F > 0 \,|\, x, K) = P((\mu_D - \mu_F - 0.7)/0.167 > -0.7/0.167 \,|\, x, K)$$
$$\approx P(\text{Normal}(0, 1) > -4.2) = 0.9999.$$

Hence the analyst is confident that the serious injury rate for mobile units is higher than for fixed installations, given the data and the model used.

The linear regression analysis can be performed similarly, but it is more technical. Again we follow the three steps:

1. Establish a probabilistic model.
2. Assign a prior distribution on the parameters of interest.
3. Use Bayes' theorem to establish the posterior distribution of the parameters.

As above, we use Poisson distributions for the number of events, but now we assume that the rate is dependent on the year and has the linear form $\mu_i = \alpha + \beta i$. By assigning prior distributions on α and β, we can derive a posterior distribution for α and β, and in particular the distribution of the rate for year 12: $\mu_{12} = \alpha + 12\beta$. Figure 6.2 shows the posterior distribution of

Table 6.1 *Serious injury data for fixed installations and mobile units.*

	Production (fixed) installations	Mobile units	All units
Number of serious injuries $\Sigma\, X_i$	359	159	508
Manhours $\Sigma\, c_i$ (in millions)	283.6	82.4	366.0
Normalised number of serious injuries μ^*	1.23	1.93	1.39

μ_{12} for fixed installations when assuming that the priori distributions of α and β are normal with mean 0 and variance 10^6 (hence we assign more or less the same prior probability for these parameters to be included in any interval of the same length). To compute the posterior density, the so-called MCMC (Markov Chain Monte Carlo) method based on the Metropolis–Hastings algorithm (Hamada *et al.*, 2008) is used, with 100 000 simulations. From the density, we obtain a 90 per cent credibility interval for μ_{12}: [0.40, 0.79]. The mean of the distribution is 0.59. The observational data completely dominate the posterior distribution.

6.2.2 Risk is defined through uncertainties

The aim of the risk assessment is now to describe the uncertainties about unknown quantities, so the first task is to identify these quantities.

The data for the years 1–11 are historical data; what we are concerned about are the future number of serious injuries and the normalised version per million manhours, and in particular these quantities (denoted X and Y) for year 12. To simplify our analysis we restrict attention to year 12.

The aim of the risk assessment is to predict these quantities and to describe uncertainties.

Table 6.1 shows the basic data from Table 5.2. From these data we predict 1.23, 1.93 and 1.39 serious injuries per million manhours for production (fixed) installations, mobile units and total, respectively, assuming no trends.

Next we need to address uncertainties, which is a key component of risk. Consider the total number of serious injuries in year 12, X, and divide the year into 365 days. For the sake of simplicity, assume that the number of manhours during day i is a fixed number d not depending on the day of the year. Hence $d365 = c_{12}$. What is the knowledge-based probability that a serious injury occurs (we denote this event A) on day 1? Based on the data observed for the previous 11 years, and assuming no trend, the assessor assigns a probability of A, P(A), equal to $d1.39/10^6$. For day two, the assessor

Table 6.2 *Serious injury data for fixed installations, for the no-trend and trend cases.*

	No trend	Trend
Predicted number of serious injuries (normalised)	37 (1.23)	19 (0.63)
90% prediction interval for number of serious injuries (normalised)	[27, 47]	[12, 26]
90% prediction interval for the normalised number of serious injuries	[0.90, 1.57]	[0.40, 0.87]

would make the same assignment, although the information basis is slightly improved. At day two the assessor can add occurrence or non-occurrence of an event on day 1. However, this additional information is negligible compared to that gained from the previous 11 years. Hence the knowledge-based probability is not changed. This argument can be repeated and we are led to an approximate binominal distribution for X, with parameters 365 and $d1.39/10^6$, which in its turn can be approximated by a Poisson distribution with mean $365d1.39/10^6 = c_{12} 1.39$.

Thus if $c_{12} = 40$, we predict $55.6 \approx 56$ events and obtain the following 90 per cent prediction interval for X using the Poisson distribution: [44, 68]. For the normalised number of serious injuries per million manhours (Y), the interval becomes: [1.1, 1.7]. The analogous rate results for fixed installations and mobile units are [27, 47] (normalised [0.90, 1.57]) and [12, 27] (normalised [1.2, 2.7]), respectively, assuming that the manhours are 30 and 10.

Next we incorporate trends, and let us again use the data for fixed installations as an illustration. Using the regression line in Figure 5.3, we obtain a prediction of the number of serious injuries rate per million manhours equal to 0.63. Then arguing as above, the uncertainty is expressed by a Poisson distribution with parameter 0.63. A 90 per cent prediction interval for the number of serious injuries then becomes (assuming 30 million manhours): [12, 26], which normalised gives the interval [0.40, 0.87]. The results differ strongly from the non-trend case, as shown in Table 6.2. For example, the prediction intervals are disjunct, [0.90, 1.57] and [0.40, 0.87], which clearly demonstrates the importance of the assumptions made.

A probabilistic analysis is always based on a set of assumptions, for example "no trend". But the assumptions could be wrong (leading to poor predictions). If there is a trend in the injury numbers it would be more reasonable to predict a rate of 0.63 than 1.23 next year. Only hindsight can show which one is the best prediction, but the above analysis makes it clear that a simple transformation of the historical figures (as reported in Table 6.1) can lead to very poor predictions.

By attempting to understand the data, by assuming a trend and carrying out a regression analysis, we may be able to improve the predictions. But we may also end up "over-interpreting" the data in the sense that we look for all sorts of explanations for why the historical figures are as they are. Perhaps injury rates are falling; perhaps the trend arrow will be reversed next year. We can analyse possible underlying conditions that can affect the injury rates, but it is not easy to reflect what the important factors are, and what is "noise" or arbitrariness.

An analysis based on the historical numbers could easily become too narrow and imply that extreme outcomes are ignored. Surprises occur from time to time, and suddenly an event could occur that dramatically changes the development, with the consequence that the rates jump up or down. In a risk analysis such events should ideally be identified. However, the problem is that we do not always have the knowledge and insights to be able to identify such events, because they are extremely unexpected.

The risk perspective adopted motivates assessment of uncertainties beyond the probabilities assigned. The practical tool used may be a list of uncertainty factors that could strongly influence the number of injuries to occur in the future. For this case, a number of such factors were identified, including:

(a) Continuous safety improvement: efforts are made to avoid accidents, in line with overall policy documents both on a company and governmental level.
(b) Economic climate: if the oil price drops we may experience a stronger focus on production and a reduced willingness to use money on safety measures.
(c) Major accidents: a large accident in the industry may give increased focus on safety and a further reduction in the number of injuries, or the result could be a shift in attention with more attention being paid to large-scale accidents and less on typical working accidents.
(d) Ageing of installations: many of the installations are old, and deterioration and ageing are increasing problems. These phenomena could lead to operational changes and possibly an increased rate of failures and incidents.

These factors may not be dramatic for the first year (year 12), but if we extend the period of analysis to several years, they could cause dramatic changes in the injury rates.

To assess the importance of these factors a qualitative crude assessment can be carried out along the following lines, as mentioned in Section 2.8 (Aven, 2008b; Flage and Aven, 2009).

Each factor's importance is measured using a sensitivity analysis. Is changing the factor important for the risk indices considered, for example the probability that the serious injury rate exceeds a specific number? If this is the case, we next address the uncertainty of this factor. Are there large uncertainties about this factor? If the uncertainties are assessed as large, the factor is given a high risk score. Hence to obtain a high score in this system, the factor must be judged as important for the risk indices considered and the factor must be subject to large uncertainties. This uncertainty assessment goes beyond the probabilistic analysis. Aspects to be considered to judge the uncertainties to be high are

- The phenomena involved are not well understood; models are non-existent or known/believed to give poor predictions.
- The assumptions made represent strong simplifications.
- Data are not available, or are unreliable.
- There is lack of agreement/consensus among experts.

For our case, let us consider the analysis when there is no trend. Then certainly this assumption ("no trend") (which is closely related to (a) represents a major uncertainty factor. It obviously scores highly on sensitivity and uncertainty and is thus an important factor. The predicted results could be poor compared to the real observations in coming years.

Next let us consider one of the other factors, the economic climate. If the economic climate changes, this may strongly affect the safety level. This was demonstrated in the late 90s when the industry faced an increased pressure to reduce costs. The industry is not likely to make the same "mistake" again but still we would judge the economic climate as important for the safety level. However, in a short-term perspective (year 12) we do not assign a high score to this uncertainty factor: we are quite confident that the economic climate will not give large changes in the injury rates, although we may experience a much lower oil price. Of course in a longer time perspective, the situation will be different.

Finally in this section we would like to discuss the appropriateness of introducing relative frequencies, or chances as we would call them in this context (see Section 2.5).

In the analysis in this section we have not introduced a stochastic model – a chance model – expressing aleatory uncertainty, i.e. variation in populations of similar units. The reason is twofold:

1. As was noted in Section 5.2 the meaning of the parameters (μ) is not clear (we need to define the average number of serious injuries when considering an infinite number of similar periods of exposure times).

2. It is not easy to see the additional insights gained in this case by introducing these models compared to the more direct and simple approach presented in this section.

However, if the analysts prefer to introduce these models, and they are justified, they can be seen also as tools for assessing the uncertainties about A and C. The assessment of the parameters of the models is then not the end product of the analysis as in the (A,C,P_f) case. Having established the posterior distribution of the parameters (μ), we use the law of total probability to establish the so-called predictive distributions of the observable quantities, for example the number of serious injuries X in one-million exposed hours (thus $c = 1$). The formula used is

$$P(X = x) = \int P(X = x|\mu)f(\mu|K)d\mu,$$

where $f(\mu \mid K)$ is the posterior distribution of μ and $P(X = x \mid \mu)$ is the Poisson distribution with parameter μ. It can be shown that X has a so-called Poisson gamma distribution (also referred to as the negative binomial distribution), i.e. (Vose, 2008):

$$P(X = x) = [\Gamma(\alpha' + x)/(\Gamma(\alpha')\Gamma(x))] \, [\beta'/(\beta' + 1)]\alpha' \, [1/(\beta' + 1)]^x, x = 0, 1, 2, \ldots,$$

where Γ is the gamma function and $\alpha' = \alpha +$ numbers of observed injuries, and $\beta' = \beta + \Sigma \ c_i$. For large α' this distribution is close to a Poisson distribution.

6.3 Case 2: QRA of LNG plant

6.3.1 Risk is defined through frequentist probabilities (chances)

We refer to Sections 4.2 and 5.3. In Section 5.3 we used models, hard data and expert judgements to estimate the parameters of interest, i.e.

p = individual risk for a specific person in the group having the highest risk, i.e. the probability that a specific person (arbitrarily chosen) shall be killed due to the activity during a period of one year, and

the f–n curve G(n) expressing the frequency f (i.e. the expected) number of accidents that lead to minimum n number of fatalities, which can also be interpreted as the probability of an accident with at least n fatalities, i.e.

$$G(n) = E_f[Y(n)],$$

where Y(n) denotes the number of accidents with at least n fatalities during a period of one year.

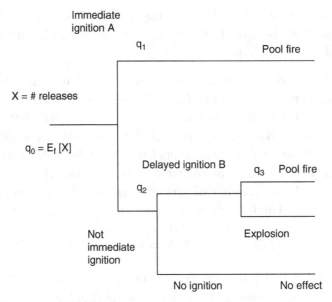

Figure 6.3   Event tree for the LNG plant case (based on Figure 4.7).

These parameters (p and G(n)) are unknown and need to be estimated and associated uncertainties assessed. We studied the point estimates in Section 5.2. Now we address the problem of assessing the uncertainties about the true value of these parameters. The tool for this purpose is knowledge-based (subjective) probabilities. The approach is referred to as the probability of frequency approach (refer to Section 6.1).

The analysis is then again an application of the Bayesian framework, which comprises the following steps:

1. Establish a probabilistic model.
2. Assign a prior distribution on the parameters of interest.
3. Use Bayes' theorem to establish the posterior distribution of the parameters.

The analysis is, however, quite different from the one presented in Section 6.2, as the analysis is based on models with many parameters and the data are rather scarce. To illustrate the analysis we will use a simplified version of the event tree presented in Figure 4.7; see Figure 6.3. We see that if a release occurs, it can either result in a pool fire, an explosion or no effect, depending on the results of the branching events, immediate ignition, and delayed ignition. The following parameters have been introduced:

$$q_0 = E_f[X]$$
$$q_1 = P_f(A)$$

$$q_2 = P_f(B \mid \text{not A})$$
$$q_3 = P_f(\text{pool fire} \mid \text{not A,B})$$

For q_1, q_2 and q_3 it is tacitly assumed that the probabilities are conditional on the occurrence of a release.

The model provides four scenarios:

S_1: release – A – pool fire
S_2: release – not A – B – flash (pool) fire
S_3: release – not A – B – explosion
S_4: release – not A and – not B – no effect.

Assume that the number of people exposed to scenario S_i is v_i, where $v_1 = 0$, $v_2 = 50$ and $v_3 = 100$. Furthermore, assume that the fraction of fatalities is d_i, where $d_2 = d_3 = 0.1$.

Let N denote the number of fatalities. Then these assumptions mean that N equals 5 in the case of scenario S_2 and N equals 10 in the case of scenario S_3, and zero otherwise.

The stochastic model used is thus described. To interpret the parameters we need to construct infinite populations of similar situations to the one studied. For example, q_1 represents the fraction of times immediate ignition occurs in the case of a release and the situation is similar to the one studied.

If we know all the parameter values we can calculate the contributions to p and G(n). Suppose as an example that

$$q_0 = E_f[X] = 0.005 \ (\approx P_f(\text{one release occurs in the period considered}))$$
$$q_1 = P_f(A) = 0.3$$
$$q_2 = P_f(B \mid \text{not A}) = 0.2$$
$$q_3 = P_f(\text{pool fire} \mid \text{not A,B}) = 0.4.$$

This gives the following probabilities for the scenarios:

$P_f(S_1 \mid \text{release}) = P_f(A) = q_1 = 0.3$
$P_f(S_2 \mid \text{release}) = P_f(\text{not A, B, pool fire}) = (1-q_1) q_2 q_3 = 0.7 \times 0.2 \times 0.4 = 0.056$
$P_f(S_3 \mid \text{release}) = P_f(\text{not A, B, explosion}) = (1-q_1) q_2 (1-q_3) = 0.7 \times 0.2 \times 0.6 = 0.084$
$P_f(S_4 \mid \text{release}) = P_f(\text{not A, not B}) = (1-q_1)(1-q_2) = 0.7 \times 0.8 = 0.56.$

By multiplying these numbers with $q_0 = E_f[X] = 0.005$, we obtain the unconditional probabilities $P(S_i)$. Hence there is a probability of 0.00042 that a release occurs and this release leads to an explosion.

It remains to include the fatality figures, 5 in the case of S_2, 10 in the case of scenario S_3, and 0 otherwise. Table 6.3 summarises the results.

Table 6.3 *Probability distribution for the number of fatalities associated with the event tree of Figure 6.2.*

N: number of fatalities associated with release as defined by event tree in Figure 6.2	P_f	E_f contribution
0	0.99930	0
5	0.00028	0.0014
10	0.00042	0.0042

From these figures we see that $P_f(N > 0) = 7 \times 10^{-4}$, i.e. the probability of a fatal accident is 0.07 per cent, and that

$$E_f N = 0.0056,$$

i.e. the expected number of fatalities due to this risk contribution is 0.0056. From Table 6.3 we have the numbers necessary to calculate the f–n curve contribution: the probability that an accident occurs with at least n fatalities. To compute an individual risk index we make a simplified assumption: 100 persons are exposed, and the IR is the same for all persons. Then

$$p = IR = E_f[X]/100 = 0.000056 = 6 \times 10^{-5}.$$

This term represents the average individual risk.

In practice we do not know the values of the parameters, and we have to replace the parameters by estimates $(q_i)^*$, $(P_f)^*$, $(E_f)^*$, etc. The above figures thus have to be interpreted as estimates of the underlying parameters. Thus we have for example $(E_f N)^* = 0.0056$.

The aim of the risk assessments in this case is to describe uncertainties, and the tool used is knowledge-based probabilities P. Next we will show how the uncertainty analysis is carried out for this case, using the event tree model.

The quantity of interest for this analysis is G(n) and p. Let us concentrate our focus on G(1), the relative frequency probability of at least one fatality, and to simplify the notation we refer to this quantity as r. From the above analysis we have established a relationship (model) between this quantity and the underlying model parameters: q_0, q_1, q_2 and q_3:

$$G(1) = r = P(S_1) + P(S_2) = q_0[(1 - q_1)q_2q_3 + (1 - q_1)q_2(1 - q_3)]$$
$$= q_0(1 - q_1)q_2.$$

The aim of the analysis is now to establish uncertainty distributions on the q_i parameters and use the event tree model to propagate these uncertainties to an uncertainty distribution for r. A numerical example will explain the ideas.

Table 6.4 *Knowledge-based probabilities for the
parameters q_0, q_1, q_2 and q_3.*

Parameter	Distribution type	Interval
q_0	Uniform	[0.003, 0.007]
q_1	Uniform	[0.2, 0.4]
q_2	Uniform	[0.1, 0.3]
q_3	Uniform	[0.1, 0.7]

Let us first consider q_0, the expected number of releases. As an estimate of q_0 we used 0.005. To reflect uncertainties we use a subjective probability distribution. This distribution may for example be a beta-distribution, a triangular distribution or a uniform distribution. For this case we will simply assume that the analysts (experts) specify a uniform distribution on the interval [0.003, 0.007], which means that the analysts (experts) are confident that the true q_0 lies in this interval, and that their degree of belief that q_0 lies in the interval [0.003, 0.005] is the same as that of [0.005, 0.007], namely 50 per cent. We make similar assumptions for the other parameters. See overview in Table 6.4.

Using these distributions and assuming "independent" distributions for the q_i parameters we can calculate the knowledge-based distributions for r. Independence here means that if for example we know that q_2 is equal to 0.12 (say), this would not affect our uncertainty assessment of q_3 (say).

To establish the output distributions using analytical formulae is difficult. It is easier to use Monte Carlo simulation, and this is the common approach for performing this type of uncertainty assessment. In this case the analysis was carried out using Matlab 2007. Random numbers for each parameter are drawn and using the formula $r = q_0 (1 - q_1) q_2$ we obtain the associated uncertainty distribution of r, shown in Table 6.5 and Figure 6.4. Note that these values are estimates of the probabilities given by the input of the Monte Carlo simulations: the uniform distributions and the formula $r = q_0 (1 - q_1) q_2$. The estimation error is small as the number of replications is large (10^7). Hence there is a knowledge-based probability of 43% that the chance of at least one fatality is in the interval (0.04%, 0.07%].

From this simple example it should be clear how to perform a similar analysis with a large number of parameters.

6.3.2 Risk is defined through uncertainties

The aim of the risk assessment is now to describe the uncertainties about unknown quantities, so we need to identify these quantities. We focus on "observable quantities", such as

Table 6.5 *Knowledge-based probabilities P for r = G(1).*

Interval for r	Interval for r. Reformulated intervals (%) (× 10⁻²)	Simulated probability
≤0.0002	≤ 0.02	0.00
(0.0002, 0.0004]	(0.02, 0.04]	0.13
(0.0004, 0.0007]	(0.04, 0.07]	0.43
(0.0007, 0.0010]	(0.07, 0.10]	0.29
(0.0010, 0.0013]	(0.10, 0.13]	0.13
(0.0013, 0.0016]	(0.13, 0.16]	0.02
> 0.0016	> 0.16	0.00

Simulated probability
distribution of r = G(1)

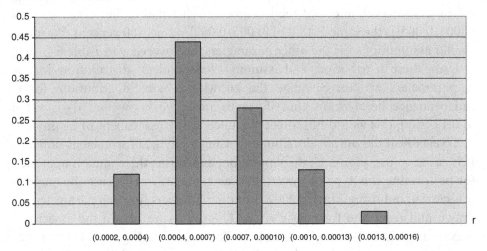

Figure 6.4   Knowledge-based probabilities P for r = G(1) based on Table 6.5.

N: the number of fatalities (third parties)

D: the occurrence of an accident leading to a fatality of person z (arbitrarily chosen).

The aim of the risk assessment is to predict these quantities and to describe uncertainties. In this case the predictions would be straightforward, no fatalities and that the event D would not occur. However, there are uncertainties and accidents could occur leading to deaths. To describe these uncertainties we use the event tree models and knowledge-based probabilities. We may introduce chances (relative frequencies), but in this case let as assume that the

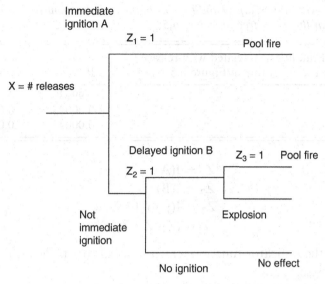

Figure 6.5 Event tree for the LNG plant case (based on Figure 4.7). No chances introduced.

analysts decide not to do so. They find that (refer to similar comments in Section 6.2 and discussion in Section 6.5):

- it is difficult to give meaningful interpretations of the chances
- the introduction of the chances makes the analysis more complex, and no added value is identified.

We note that if such an analysis were to be carried out it would take the form studied in the previous section (Section 6.3.1). However, the end product would be unconditional uncertainty distributions (predictive distributions) of N and D obtained by using the law of total probability, as was indicated also in Section 6.2.2 for the injury case. The formula used is

$$P(N = n) = \int P(N = n \mid q)f(q)dq, \tag{6.1}$$

where q is the vector of parameters and f(q) is the assigned uncertainty distribution of q (in the above case independent uniform distributions over each parameter).

   In the following we analyse the system without introducing chances, and we restrict attention to N. To ease the assignment of knowledge-based probabilities of N we introduce a model g, an event tree model; see Figure 6.5. Here

   X = Number of releases (which is equal to 1 if a release occurs and 0 otherwise as we ignore the probability of two releases in the period studied)

Table 6.6 *Knowledge-based probability distribution for the number of fatalities associated with the event tree of Figure 6.5.*

N: number of fatalities associated with release as defined by the event tree in Figure 6.5	P	E contribution
0	0.99930	0
5	0.00028	0.0014
10	0.00042	0.0042

$$Z_1 = I(A)$$
$$Z_2 = I(B)$$
$$Z_3 = I(\text{pool fire}).$$
$$Z = (Z_1, Z_2, Z_3)$$

Here I is the indicator function which is equal to 1 if the argument is true and 0 otherwise.

The model expresses that

$$N = g(X, Z) = 5X(1 - Z_1)Z_2Z_3 + 10X(1 - Z_1)Z_2(1 - Z_3),$$

as the number of fatalities is 5 in the case of scenario 2 and this scenario occurs if $X(1-Z_1)\, Z_2\, Z_3 = 1$, and the number of fatalities is 10 in the case of scenario 3 and this scenario occurs if $X(1-Z_1)\, Z_2\, (1-Z_3) = 1$.

The quantities X and Z are unknown, and knowledge-based probabilities are used to express the uncertainties (degree of belief). Suppose the following assignments have been made given the background knowledge K of the analysts (experts):

$$P(X = 1) = EX = 0.005$$
$$P(Z_1 = 1) = P(A) = 0.3$$
$$P(Z_2 = 1 \mid Z_1 = 0) = P(B \mid \text{not A}) = 0.2$$
$$P(Z_3 = 1 \mid Z_1 = 0, Z_2 = 1) = P(\text{pool fire} \mid \text{not A, B}) = 0.4.$$

We have used the same numbers as the estimates of the relative frequency-interpreted probabilities in Section 6.3.1. However, the meaning of the numbers is completely different. For example, $P(Z_1 = 1) = P(A) = P(A \mid K) = 0.3$ means that the analysts (experts) consider the uncertainty of immediate ignition occurring (given a release) as the same as drawing a red ball out of an urn which comprises ten balls and three are red.

To compute the distribution of N given this input we follow the rules of probability as in the previous section. The results are shown in Table 6.6. We see that the numbers are the same as in Figure 6.5, as the input probabilities are the same. As stressed in the previous paragraph the interpretation is, however, different.

The probabilities assigned are based on a background knowledge K, which includes a number of assumptions. Here are some few examples, evident from the above analysis

- the event tree model
- a specific number of exposed people
- a specific fraction of fatalities in different scenarios.

As examples of other assumptions made in this case, not revealed, however, by the above analysis, we mention:

- The probabilities and frequencies of leakages were based on a database for offshore hydrocarbon releases.
- All vessels and piping are protected by water application like monitors, hydrants.
- Release rates are constant throughout the release duration time.

The understanding of the physical phenomena and the computer codes used also strongly affects the results.

Vinnem (2010) illustrates the dependencies of the assumptions made by pointing out that the frequency of accidents with at least 100 fatalities increased by a factor of 56 when compared to the results from the initial risk assessment performed for the operator of the plant. The initial assessment was performed before engineering studies had started, whereas the updated study made by the engineering contractor reflected all the engineering details.

Vinnem (2010) also points to the assumption made in this case that, in the event of impact of a passing vessel on an LNG tanker loading at the quay, the gas release would be ignited immediately, presumably by sparks generated by the collision itself. However, according to Vinnem (2010), no explanation was provided of how such ignition of a very heavy and cold gas could occur physically. He concludes that it is very hard to foresee how it could be caused in this way. The implications of the assumption are important for the further analysis:

However, the implication of this assumption was that it was unnecessary to consider in the studies any spreading of the gas cloud due to wind and heating of the liquefied gas, with obvious consequences for the scenarios the public might be exposed to. Such a very critical assumption should at least have been subjected to a sensitivity study in order to illustrate how changes in the assumption would affect the results, and the robustness of the assumption discussed. None of this, however, has been provided in any of the studies (Vinnem 2010).

A long list of uncertainty factors could be generated by looking at the many assumptions made in the risk assessment. However, for the purpose of the present analysis, the above points are sufficient.

6.4 Case 3: Design of a safety system

The assessment in this case would be quite similar to those presented for the two previous cases, and to avoid tedious repetitions we will focus on some specific issues only:

the use of the models for computing the MFDT:
$\lambda\tau/2$ (alternative I) and $(\lambda\tau)^2/3$ (alternative II).

The standard analysis to describe the uncertainties is to use the Bayesian framework, which we know consists of the following basic steps:

1. Establish a probabilistic model.
2. Assign a prior distribution on the parameters of interest.
3. Use Bayes' theorem to establish the posterior distribution of the parameters.

The model in this case was introduced in Section 5.4. Our focus now is the parameters λ and MFDT. As a prior distribution on λ we may use a gamma distribution f with parameters α and β and we can proceed as in Section 6.2 to obtain the posterior distribution $f(\lambda|$ data). From this distribution we can derive the posterior distribution for MFDT. We introduce $\gamma = $ MFDT and obtain

$$P(\gamma < x) = P(\lambda\tau/2 < x) = P(\lambda < 2x/\tau) \quad \text{for alternative I, and}$$
$$P(\gamma < x) = P((\lambda\tau)^2/3 < x) = P(\lambda < (3x)^{1/2}/\tau) \quad \text{for alternative II.}$$

Instead of assessing uncertainties about the parameters λ and γ, we may instead focus on observable quantities such as Y_t defined as the downtime of the system in the interval $[0, t]$. Provided that λ is known, we know from renewal reward theory (see e.g. Aven and Jensen (1999), p. 250) that Y_t/t is approximately normally distributed with mean equal to MFDT and variance equal to Var(Y)/$(t\tau)$ where Y is the time the system is down in the first cycle, i.e. $[0, \tau]$. However, this normal distribution just reflects aleatory uncertainties (random variation), not epistemic uncertainties. To include the latter types of uncertainties, a distribution also has to be specified for λ. Combining these distributions, we obtain the predictive distribution for Y_t similar to (6.1).

6.5 Discussion

In this section we will discuss the scientific quality of the risk assessment when the objective is uncertainty description, again using the reliability and validity requirements (refer to Section 3.3). The three examples above will be used to illustrate the analysis. The validity criterion V1 (that the produced risk numbers are accurate compared to the underlying true risk) is not relevant in this case.

6.5.1 Validity criteria V2–V4

The validity criteria relate to the degree to which the risk assessment describes the specific concepts that one is attempting to describe. The specific criteria V2–V4 state:

- The degree to which the assigned probabilities adequately describe the assessor's uncertainties of the unknown quantities considered (V2).
- The degree to which the epistemic uncertainty assessments are complete (V3).
- The degree to which the analysis addresses the right quantities (V4).

Concerning criterion V4, the degree to which the analysis addresses the right quantities, we refer to Section 5.5.2.

It is not straightforward to verify that the validity requirement V2 is met, and there is an ongoing research and discussion in the literature addressing this issue. It is outside the scope of this discussion to give a full account of this research and discussion, but we will highlight some important principles and procedures (refer Aven, 2003, 2004; Cooke, 1991; Lindley *et al.*, 1979):

 (i) Coherent uncertainty assessments are achieved by using the rules of probability, including Bayes' theorem for updating of assessments in the case of new information. See Appendix A.

 (ii) Comparisons are made with relevant observed relative frequencies if available. For example, if history shows that out of a population of 1000 units, two have failed, we can compare our probability to the rate 2/1000.

(iii) Training in probability assignments is required to make assessors aware of heuristics as well as other problems of quantifying probabilities such as superficiality and imprecision (which relate to the assessor's possible lack of feeling for numerical values). Heuristics for assigning probabilities are easy and intuitive ways to deal with uncertain situations. The result of using such heuristics is often that the assessor unconsciously tends to put "too much" weight on insignificant factors. An example of a heuristic is the so-called availability heuristic (Kahneman and Tversky, 1974):

 The assessor tends to base his probability assignment on the ease with which similar events can be retrieved from memory. The occurrence of events where the assessor can easily retrieve similar events from memory is likely to be given higher probabilities than the occurrence of events that are less vivid and/or completely unknown to the expert. A classic example is a person who argues that cigarette smoking is not unhealthy because his grandfather smoked three packs of cigarettes a day and lived to be 95.

Heuristics also need to be given attention when professional analysts and experts assign probabilities, but this is mainly a problem when lay people assign probabilities.

(iv) Using models, including probability models, to simplify the assignment process.

(v) Using procedures for incorporating expert judgements.

(vi) Accountability: the basis for all probability assignments must be identified.

All of these areas are important but with the exception of (iv), we just briefly address them here. Item (iv) is in fact thoroughly covered throughout this book; see for example discussions in Section 5.5 and below concerning model uncertainties.

One more issue should be mentioned: a risk analyst (an expert) may assign a probability that completely or partially reflects inappropriate motives rather than his deeply felt belief regarding a specific event's outcome. As an example, it is hard to believe that a sales representative on commission would make a completely unprejudiced judgement of two safety valves, one of which belongs to a competitor's firm. Another example is an engineer that has been involved in the design process and later is asked to judge the probability of failure of an item he personally recommended to be installed. The engineer claims that the item is "absolutely safe" and assigns a very low failure probability. The management may reject the sales representative's judgement without much consideration since they believe that inappropriate motives have influenced his judgement. The engineer's judgement might not be rejected just as easily since he obviously is a company expert in this area. On the other hand, incentives are present that might affect his probability specification.

Motivational aspects will always be an important part of evaluating probabilities. In general we should be aware of the existence of incentives that in some cases could significantly affect the assignments. However, we will conclude that motivational aspects are not a problem when professionals perform risk assessments. On the contrary, in general professional analysts (experts) would not, by intention, perform a biased assessment, influenced by motivational factors. Their jobs would not last long if their reputation were questioned. However, their approach to the assessment and the methods used could be strongly in favour of one specific party. For example, when performing a standard risk analysis of a process plant, one may argue that important uncertainty factors are camouflaged, and hence V3 is not met. See discussion below. Do the analysts do anything about this? Do they report on this? Probably not, as it is not in the interest of the client (plant operator). Thus indirectly, motivational aspects are an important issue when assessing the results of risk assessments.

These principles and procedures provide a basis for establishing a standard for the probability assignments; the aim being to extract (elicit) and summarise knowledge about the unknown quantities (parameters), using models, observed data and expert opinions. It seems reasonable to say that the requirement V2 is met provided that this standard is followed.

Next we address V3 and let us first focus on the probability of frequency approach. Following this approach, the analysts (experts) are to express the epistemic uncertainties about the parameters of the probability models using knowledge-based probabilities. In practice it is difficult to perform a complete uncertainty analysis within this setting, i.e. fulfil requirement V3. In theory an uncertainty distribution on the total model and parameter space should be established, which is impossible to do. So in applications only a few marginal distributions on some selected parameters are normally specified, and therefore the uncertainty distributions on the output probabilities are just reflecting some aspects of the uncertainty. This makes it difficult to interpret the produced uncertainties. This problem was most relevant for the LNG example where the assessment is based on complex models with hundreds of parameters. The challenges of meeting the V3 criterion also applies to the (A,C,U) perspective, although they are considered smaller due to the inclusion of uncertainty factors. The reduced use of probability models and parameters also means that V3 is not so difficult to meet as for the probability of frequency approach.

All approaches are based on the use of knowledge-based probabilities, and these reflect the uncertainties (degree of belief) of the assessors conditional on the background knowledge K. As shown by the examples in Section 6.2.2 and 6.3.2, these probabilities may in some cases camouflage uncertainties if not addressed. The assigned probabilities are conditioned on a number of assumptions and suppositions. Uncertainties are often hidden in the background knowledge, and we may consequently question whether the assigned probabilities are able to adequately describe the uncertainties (V3, partly V2).

As an example, think of the assumption made in the LNG case that, in the event of impact of a passing vessel on an LNG tanker loading at the quay, the gas release would be ignited immediately, presumably by sparks generated by the collision itself. This assumption could be wrong. Uncertainties are not revealed when not assessing the uncertainties about this assumption.

This issue is discussed by for example Mosleh and Bier (1996). They refer to a subjective probability $P(A|X)$ which expresses the probability of the event A given a set of conditions X. As X is uncertain (it is a random variable), a probability distribution for the quantity $h(X) = P(A|X)$ can be constructed. Thus there is uncertainty about the random probability $P(A|X)$. However, we will stress that the probability is not an unknown quantity (random variable) for the analyst (expert). To make this clear, let us summarise the setting of subjective probabilities. A subjective probability $P(A|K)$ is conditional on the background knowledge K, and some aspects of this K can be related to X as described by Mosleh and Bier (1996). The analyst (expert) assigns his/her probability based on K. If he/she finds that the uncertainty about X should be reflected, he/she would adjust the assigned probability using the law of total probability. This does not mean, however, that $P(A|K)$ is uncertain, as such a statement would presume that a true probability value exists. The assessor needs to clarify what is uncertain and subject to the uncertainty assessment, and what constitutes the background knowledge. This is a key point to meeting the criterion V2, but also V3. From a theoretical point of view one may think that it is possible (and desirable) to remove all such Xs from K, but in a practical risk assessment context that is impossible. We will always base our probabilities on some type of background knowledge, and often this knowledge would not be possible to specify using quantities such as X.

We conclude that the assessment fails to meet V3 in this case. However, for the two other cases the assessments meet the criterion when restricting attention to the parameter uncertainties. Model inaccuracies (uncertainties) are not incorporated. Let us look more closely into this latter type of uncertainties. We will argue that the epistemic uncertainty analysis cannot and should not aim at quantifying the model inaccuracies, i.e. to meet the criterion V3 these uncertainties are not relevant.

Model uncertainties

We use the safety design system alternative II as an illustrating example. Let G denote the model used for the MFDT, i.e. $G(\lambda) = (\lambda\tau)^2/3$. This model is based on independence between the two components when λ is known. The model inaccuracy is defined by the difference between the "true" MFDT and the model output, i.e. $MFDT - G(\lambda)$. This difference is also referred to as model uncertainty; see e.g. Östergaard *et al.* (1996), Kaminski *et al.* (2008)

and Nilsen and Aven (2003). It obviously needs to be addressed as the uncertainty assessments are conditional on the use of this model. But how should we deal with this "error" – should we quantify it?

No, is our clear answer (Aven 2010b). It is not meaningful to quantify the model inaccuracy. The point we make is that if the model is not considered good enough for its purpose, it should be improved. The uncertainty assessments are based on the model used. Of course, when observations of MFDT (i.e. of Y_t/t for large t) are available, we would compare the assessments of MFDT which are conditional on the use of the model G, with these observations. The result of such a comparison provides a basis for improving the model and accepting it for use. But at a certain stage we accept the model and apply it for comparing options and making judgements about, for example, risk acceptance (tolerability). Then it has no meaning in quantifying the model inaccuracy. The results are conditional on the model used. Instead of specifying $P(MFDT \leq x)$ directly, we compute $P(G(\lambda) \leq x \mid K)$ and G is a part of the background knowledge K.

An important task for the scientific communities in different areas is to develop good models. The models are justified by reference to established theories and laws explaining the phenomena studied, and the results of extensive testing. The performance of a model must, however, always be seen in light of the purpose of the analysis. A crude model can be preferred instead of a more accurate model in some situations if the model is simpler and it is able to capture the essential features of the system performance.

In the literature, attempts have been made to explicitly incorporate the model inaccuracies (an example is given in Aven (2003) taken from the field of structural reliability analysis (SRA)). The use of $G(\lambda) = (\lambda\tau)^2/3$ means a simplification, and the idea is then to introduce an error term a, say, such that we obtain a new model $G_0(\lambda) = a(\lambda\tau)^2/3$, where a is a correction term. Clearly, this may give a better model, a more accurate description of the world. However, it would probably not be chosen in a practical case as it may complicate the assessments. It may be much more difficult to specify a probability distribution for (a, λ) than for λ. There might be lack of relevant data to support the uncertainty analysis of a, and there could be dependencies between a and λ. We have to balance the need for accuracy and simplicity.

The MFDT is also based on the use of the exponential lifetime distribution. This distribution $G(t \mid \lambda)$ is a model of the true distribution F, where λ is the parameter of the distribution, the failure rate. Model inaccuracy expresses the difference between this model relative to the true distribution. The parameter

λ we interpret as the inverse of the average lifetime for an infinite population of similar units. Epistemic uncertainties about λ are incorporated into the uncertainty analysis as shown in Section 6.4, but model inaccuracy related to the choice of probability distribution is not. We have introduced the exponential distribution to simplify the problem. If we had considered the space of all distribution functions, the assignment process would not be feasible in practice. If we are not satisfied with the exponential distribution class, we should change this, by for example using a Weibull distribution. But when running the analysis and computing the distribution of MFDT, we accept the use of a specific model which constitutes a part of the background knowledge that the assessments are based on.

The assumptions supporting a model can give rise to uncertainty factors as discussed above. For example in the case $G(\lambda) = (\lambda\tau)^2/3$ these factors could relate to independence and deterioration over time.

In the literature, various methods have been suggested to reflect model uncertainties (see e.g. Apostolakis, 1990; Nilsen and Aven, 2003; Devooght, 1998; Zio and Apostolakis, 1996). Above we briefly looked into one typical approach (the standard SRA approach). As another typical approach we refer to Apostolakis (1990) which addresses the issue of weighing different models: let M_1 and M_2 be two alternative models to be used for assigning the probability of the event A. Conditional on M_i, we have an assignment $P(A \mid K_i)$. Unconditionally, this gives

$$P(A \mid K) = P(A \mid K_1)p_1 + P(A \mid K_2)p_2, \qquad (6.2)$$

where p_i is the analyst's (expert's) subjective probability that the ith model, i.e. the set of associated assumptions, is true (here $p_1 + p_2 = 1$). In a practical decision-making context the analysts would most likely present separate assignments for the different models – $P(A \mid K_i)$, in addition to the weighted probability assignment (6.2). To specify the knowledge-based probability $P(A \mid K)$, the analysts may also choose to apply the assignment procedure given by (6.2) when p_i cannot be interpreted as a probability that a specific assumption is true. In such a case p_i must be interpreted as a weight reflecting the confidence in the model *i* for making accurate predictions.

Hence model uncertainty quantification in the sense of model weighing can be covered by the uncertainty assessment. Model weighing is a completely different issue from quantification of model inaccuracy. As stressed above, when computing $P(\text{MFDT} \leq x)$ etc., we may accept the use of specific models and procedures for weighing the models. The models and procedures are part of the background knowledge K.

6.5.2 Reliability criteria

Now let us look at the reliability criterion R: the extent to which the risk analysis yields the same results when repeating the analysis. One may expect that following the standard for probability assignments (i.e. meeting V2), would ensure that the reliability requirement R is met. However, the background knowledge that the assignments are based on need not be exactly the same from analysis to analysis. Hence we would experience differences in the probability assignments, but the differences are not likely to be large if V2 is met. This applies to R1 and R2. The criterion R3 (the degree to which the risk analysis produces identical (similar) results when conducted by different analysis teams with the same analysis scope and objectives, but no restrictions on methods and data) would in general not be met, as the background information would be different from analysis to analysis, and often this difference could be very large due to different levels of competence, research schools, tools available etc. The problem relates to the LNG case in particular as the assessments are based on many subjective judgements and assumptions. See the benchmarking exercise reported in Lauridsen *et al.* (2001) which illustrates this problem of lack of reliability (R3). For the Cases 1 and 3 we would not expect large differences from analysis to analysis as the problems are less complex and more established methods for these situations exist.

We may question the appropriateness of the reliability criteria in this setting. Obviously we would require the results not to depend on the person running the computer calculations etc., but it should not be an objective to strive for identical results for different analysis teams. According to V2, the aim is to assess uncertainties using knowledge-based probabilities. The background information for these assignments could be different from analysis to analysis, and often this difference could be very large as mentioned in the previous paragraph. Reflecting these differences may be considered an important aim of the analysis.

6.5.3 Summary of assessments and final remarks

Table 6.7 summarises the conclusions from the previous section.

Thus the reliability and validity criteria are to a large extent met, when the assessments are properly conducted. However, several problems have been identified.

For the probability of frequency approach, the validity requirement V1 is not in general satisfied, and V2–V4 are questioned:

- important uncertainty factors may be hidden in the background knowledge (V2,V3)

Table 6.7 *Summary of reliability and validity analysis. Y indicates that the criterion is met, N that it is not met and Y/N that it is met under certain conditions. The boxes are empty in cases where the criterion is not relevant.*

Approach	Criterion								
	R	R1	R2	R3	V	V1	V2	V3	V4
Probability of frequency approach $(A,C,P(P_f),K)$	Y/N	Y/N	Y/N	N	Y/N	–	Y/N	Y/N	Y/N
(A,C,U,P,K) approach	Y/N	Y/N	Y/N	N	Y/N	–	Y/N	Y/N	Y

- the uncertainty assessments may not be complete (V3)
- the analysis focus is on fictional quantities (V4).

When we ignore the hidden uncertainties in the background knowledge, the probability of frequency approach in general may meet the validity require-ment V2 if the analysis is based on a set of standards established for such assignments.

For the reliability criteria, a main problem (again within the context of the probability of frequency approach), is the fact that the background knowledge that the assignments are based on would not be exactly the same from analysis to analysis. However, if the methods and data are fixed, the differences from one analysis to another are not likely to be large if V2 is met.

As for the probability of frequency approach, the validity requirements V2 and V3 are also questioned for the (A,C,U) approach:

- important uncertainty factors may be hidden in the background knowledge (V2,V3)
- the uncertainty assessments may not be complete (V3).

The (A,C,U) approach faces the same types of reliability problems as the probability of frequency approach.

The above analysis has restricted attention to the probability of frequency approach when the risk perspective is (A,C,P_f). Alternatively we could have looked at the pure statistical approach with the use of confidence intervals. However, it is clear that this approach in general fails to meet the validity criterion as uncertainties are not described beyond the data variation expressed by the confidence intervals (provided such intervals can be estab-lished). For an analysis of the reliability criteria, we refer to Section 5.5.3.

7

Risk management and communication issues

As stressed in Chapter 2, risk assessments provide decision support on the choice of measures and arrangements. Such decisions are risk-informed, not risk-based. There is always a need for seeing beyond the results of the risk assessments. There are two main reasons for this:

1. the limitations and boundaries of the risk assessments
2. the need for taking into account other concerns than risk when making such decisions.

The analysis in the two last chapters has addressed the first point, and in this chapter we will discuss the implications of the findings for risk management and communication where point 2 is a key issue. More specifically we focus on the following topics in this chapter:

- the use of predefined risk criteria
- the use of the ALARP principle and cost–benefit type of analyses
- the role of the cautionary and precautionary principles
- risk communication
- the content and purpose of managerial review and judgement.

7.1 The use of predefined risk criteria

It is a common approach to risk management to use predefined risk criteria of the form "risk is unacceptable (intolerable) if $P > p_0$", where P is a risk-related probability index and p_0 is a fixed number. Several examples were defined in the LNG case:

IR (probability that a fixed arbitrary third person is killed due to an accident in one year) $> 1 \times 10^{-5}$

f-n curve $G(n) > g(n)$, for some n, where g is defined in Figure 3.7 ($\log g(n) = -2 + \log n$).

We will look more closely into this practice in view of the different perspectives analysed in Chapters 5 and 6, and the findings related to scientific quality.

If the scientific basis of the risk assessment is the (A,C,P_f) perspective, these criteria are all based on unknown frequentist probabilities which need to be estimated. In the examples mentioned we obtain estimates IR* and G(n)* which are to be compared to the specified criteria. However, as was noted in Section 5.5, the scientific quality requirements of validity and reliability are not in general met if the aim is accurate risk estimations, unless large amounts of relevant data exist, and for typical quantitative risk assessment such data are not available. Hence we cannot obtain confidence that the estimates are close to the "true" P_f values and the approach collapses. It is of course possible to use the criteria based on the estimates, but a strong element of arbitrariness has then been introduced, and from a scientific point of view it cannot be justified.

A conservative policy could be to replace the risk estimates by some upper values which the analysts are confident that the true risk indices are not exceeding. Hence if these upper values are below the criteria, risk is not considered unacceptable or intolerable. Such a policy would, however, also lead to unacceptability for risk estimates not falling in the intolerable region. The policy could consequently be very expensive and not attractive from a cost–benefit point of view.

In the case of the working accidents, a substantial amount of relevant data exists. Although no risk criteria were defined for this case, we may easily construct some, for example that the total serious injury rate in the coming three years (μ for years 12–14) should be less than, say 1.0. From the historical data we may then conclude that this criterion is met if we assume a trend. For the non-trend situation the criterion is not met. The estimation precision (as expressed by the confidence intervals) is reflected in these conclusions.

We see that such a risk criterion can be justified in this case. In practice it is not so common to use such criteria in situations like this. It is more common to define targets and goals, expressing for example that the observable rate should be below 0.5 in the coming years, i.e. the number of serious injuries per million manhours should be below 0.5. These targets and goals are not risk criteria like IR and the f-n curve. The target/goal of 0.5 is checked by observing the actual number of serious injuries. The rates become known and we can verify whether the targets are met or not. Then we can decide if there is a need for action and measures. This is a reactive strategy in the sense that measures are taken following the serious injuries. The problem of expressing risk is, however, avoided.

It is also common to refer to the zero-vision: the long-term goal is to completely avoid these types of injuries. Such a vision could be useful for keeping a constant focus on safety improvement in operations; however, it does not provide much help in guiding the decision-makers on the use of resources to reduce risk.

If the scientific basis is the (A,C,U) perspective, the above criteria (IR and f-n curve) are directly related to the analysts' (experts') knowledge-based probabilities. For example, the IR is considered too high if the assigned probability that a third person is killed due to an accident in one year is greater than 1×10^{-5}. Methods and procedures are established for the calculations and if the output probability exceeds 1×10^{-5} risk-reducing measures are required. However, it is acknowledged that the computed probability is conditional on many assumptions and these could hide uncertainties. A mechanical procedure for unacceptability or not cannot be justified.

Consequently, if risk criteria of the form of IR and the f-n curve are introduced, they have to be used as nothing more than reference levels to inform the decision-maker, not to provide a mechanical procedure for what is an acceptable or not acceptable risk.

These conclusions would also apply to the more detailed system level, for example in Case 3 where we study the performance of a safety system. However, many analysts would argue that there is a need for such rules in practice at this level to ease the engineering work in a development project. In an early design phase it is not feasible to specify all possible arrangements and measures in detail and perform optimisation of all attributes (feasibility, risk, cost, etc.). We must use some sort of performance characterisation. Typically, these will be industry standards, established practice and descriptions of the performance of the system, given by features such as reliability, effectiveness (capacity) and robustness. In other words, instead of specifying in an accurate way, what system we need, we specify the performance of the system. As an example consider the requirement that

The MFDT of the safety system in Case 3 must be below 0.05.

The starting point for choosing a specific requirement could be historical data, standards, or the desire to achieve a specific risk level or improvement. The engineering process will produce a system layout that should meet this requirement. Just looking at the findings of the risk assessment would exclude alternative I. Only alternative II will meet this criterion.

However, for the 5% requirement to be meaningful it must not be seen as a sharp line; we should always look for alternatives and then evaluate their performance. Whether the analysis team calculates a reliability of 3%, 7% or

4% is not so important – depending on the situation we may accept all these levels. The interesting question is how the alternatives perform relatively, concerning reliability, costs and other factors. The figure of 5% must be seen as a starting point for further optimisation.

Instead of a sharp level, ranges may also be used, such as the categorisation used for Safety Integrity Level (SIL) requirements, in accordance with IEC 61511, for example a failure probability in the range 10%–1%.

This reasoning is supported by the lack of precision in the risk assessment as discussed in previous chapters. But equally important is the need for taking into account other concerns than risk in the decision-making (point 2 mentioned in the introduction of this chapter). We may require that the MFDT is below 0.05, but a level of 0.1 could be more than enough if other measures are implemented. The result could be improved safety and reduced costs. This is what risk management is all about. Risk criteria need to be used with care, to avoid sub optimisation. A too strong requirement regime may limit the creativity and drive for identifying the best overall arrangements and measures.

Nonetheless, many safety regulators have found the use of such criteria important for ensuring a minimum safety level. The idea is to state a minimum standard, irrespective of economy and other conditions, below which nobody is allowed to operate. An example is the approach taken to the design of platforms for environmental loads. If we go back to around 1960, the common design approach for offshore Gulf of Mexico was to design for waves with 25 year return periods. One year in the early 1960s, there were many severe storms, leading to more than a dozen platforms being toppled over due to wave overload. It was then decided that dimensioning wave load should be increased to 100 year return period in order to increase the minimum standard. One hundred year return periods are still used in the North Sea and other Norwegian waters as the minimum design wave load, without significant damage to the structure. In the last 15–20 years, it has also been required in the Norwegian operations that the installations shall withstand waves with 10 000 year return period, but then substantial damage to the structure is acceptable (Aven *et al.*, 2006).

There is not much discussion about the usefulness of such criteria. However, to implement such criteria in practice we need to understand the scientific constraints and limitations of the tool used to verify the criteria, i.e. the risk assessments. The analysis in the previous chapters has shown that we need to use the criteria as guidelines more than strict limits.

For some reflections on the ethical justification of risk acceptance criteria, see Aven (2007b) and Ersdal and Aven (2008).

7.2 The use of the ALARP principle and cost–benefit type of analyses

The ALARP principle expresses that risk should be reduced to a level that is as low as reasonably practicable. A risk-reducing measure should be implemented provided it cannot be demonstrated that the costs are grossly disproportionate relative to the gains obtained. The key point is that *risk* should be reduced; but as we have seen in earlier chapters, there are many perspectives on risk and this would also mean different perspectives on how to understand and implement the ALARP principle.

If the (A,C,P_f) risk perspective is adopted risk reduction means reduction of the probabilities (chances) P_f of undesirable events and consequences, whereas if the (A,C,U) perspective is adopted risk reduction also means uncertainty reductions. We first look more closely into the former type of perspective, and let us again use the LNG case as an example.

A measure is suggested to reduce the risk expressed by IR and the f-n curve. The estimated differences in risk reduction are calculated and by comparing these estimates with the cost of implementing the measure a basis has been established for making a judgement about gross disproportion. If the costs are large the issue quickly becomes a management issue. The need then arises for some guidelines for making judgements about gross disproportion. Cost-effectiveness indices and cost–benefit type of analyses represent the most common methods used. Of the cost-effectiveness indices the implied cost of averting a fatality (ICAF) is most relevant in this case. The index is based on a computation of the expected number of saved lives by implementation of the risk-reducing measure. Then this number is compared with the expected cost to produce the index:

$$ICAF = E[cost]/E[number\ of\ lives\ saved].$$

As the risk perspective is (A,C,P_f) we have to read this as

$$ICAF = E_f[cost]/E_f[number\ of\ lives\ saved],$$

i.e. the expected values are the means of frequentist probability distributions. These distributions and expected values are unknown, leading to estimates

$$ICAF^* = E_f^*[cost]/E_f^*[number\ of\ lives\ saved].$$

The uncertainties in the costs are often small, but the uncertainties related to the estimates of the expected number of saved lives could be large. To reflect the uncertainties we could replace the best estimate $E_f^*[number\ of\ lives\ saved]$ by a somewhat more optimistic estimate, for example the 75 per cent quantile of the epistemic distribution of $E_f[number\ of\ lives\ saved]$. To illustrate this numerically, suppose that $E_f^*[cost] = €1.0$ (million) and $E_f^*[number\ of\ lives$

saved] = 0.1. Then ICAF* = €1.0/0.1 = €10 and costs are typically considered in gross disproportion to the benefits gained. The costs are simply too high compared with the risk reduction. To reflect uncertainties we replace E_f*[number of lives saved] by the more optimistic estimate 0.5, which is judged to be the 75 per cent quantile of the epistemic distribution of E_f[number of lives saved]. This changes the computed ICAF to €2 and in most cases gross disproportion would not have been demonstrated.

Alternatively, and this is more common, we may change the criterion for what is grossly disproportional. If a limit is set to €2 (million), one may argue that the criterion should be increased to say five to take into account uncertainties. This argument is used by the HSE in the UK to motivate an increased limit in the oil and gas industry.

HSE (2001) defines an ICAF limit equal to £1 million. However, for the offshore industry an ICAF of £6 million is considered to be the minimum level, i.e. a proportion factor of six (HSE, 2006). This value is used in an ALARP context, and defines what is judged as "grossly disproportionate". Use of the proportion factor six is said to account for the potential for multiple fatalities as well as uncertainties.

However, the ICAF is still based on expected values. This strategy of using a specific adjustment factor (here six) should therefore be used with care (Aven and Abrahamsen, 2007). Think of two risk-reducing measures (a) and (b), which both cost 10 and reduce the expected number of fatalities by one (thus the ICAF is 10):

(a) Small uncertainties: The expected number of fatalities is a good prediction of the actual number of fatalities saved.
(b) Large uncertainties: The actual number of fatalities could deviate strongly from 1.

In this case with a criterion of six, none of the measures would be justified. The uncertainties in case (b) could be extreme but the level of uncertainties is not explicitly reflected in the strategy. If the expected number of saved lives is not changed, the ICAF is not changed, and gross disproportion has been demonstrated. The use of expected values as in the ICAF fails to capture important aspects of risk, in particular the probability of a large-scale event.

If C is the number of fatalities, an expected value would be an informative measure if this value is approximately equal to C, i.e. EC ≈ C. But since C is unknown at the time of the assessment, how can we be sure that this approximation would be accurate? Can the law of large numbers be applied, expressing that the mean of independent identically distributed random variables converges into the expected value when the number of variable increases to infinity? (Aven, 2009a). See also Appendix A.1.3.

Yes, it is likely that if C is the sum of a number of projects, or some average number, our expected value could be a good prediction of C. Take for example the number of fatalities in traffic in a specific country in one year. From previous years we have data that can be used to accurately predict the number of fatalities next year (C). However, in many cases the variations are much larger. Looking at the number of fatalities in Norway caused by terrorist attacks in the next year, the historical data would give a poor basis. We may assign an EC but obviously EC could be far away from C. The accuracy increases when we extend the population of interest. If we look at one unit (e.g. country) in isolation, the C numbers are in general more uncertain than if we consider many units (e.g. countries). Yet, there will always be variation/uncertainties, and in the case of extreme events, we will need to see beyond the expected value.

The literature includes a number of attempts to modify the expected value-based approaches to reflect risk aversion (which means that we dislike negative consequences so much that these are given more weight than is justified by reference to the expected value) (Levy and Sarnat, 1994). It is acknowledged that we need to take into account risks and uncertainties, and see beyond the computed expected values. However, there are "a million ways" of extending the traditional approach based on the expected net present value or the ICAF. How should we determine what is the correct or best modification? There needs to be a rationale supporting the approach.

But such a rationale is difficult to find; see Aven and Flage (2009) who review and discuss alternative approaches. Many of the extended approaches have a strong element of arbitrariness in the way they are defined, so care has to be shown when using these approaches. The problem is that the approaches fail to acknowledge that caution and precaution in cases of uncertainty cannot be captured by a probabilistic approach alone. We refer to the discussions in Sections 7.3 and 7.5.

Now let us consider the (A,C,U) risk perspective. Risk here includes uncertainty as a main component and risk reduction thus has to also cover uncertainty reduction. Consequently a policy based on ICAF and other expected value-based approaches cannot be used alone to determine ALARP. Processes that extend beyond the expected value-based and the probability-based approaches are required. Case (b) in the example above could be justified due to large uncertainties. The analysis in Chapters 5 and 6 clearly demonstrates the need for seeing beyond the computed probabilities and expected values for verifying ALARP and gross disproportion. For example in the LNG case the many uncertainty factors could justify risk-reducing measures even with rather high ICAF values. We need to implement "non-mechanical" policies that have a strong involvement of management processes as will be discussed further in coming sections. See also the approach summarised in Figure 1.4 which is developed to meet this challenge. In the following we will also justify this approach from a theoretical welfare economic point of view.

7.2.1 *Welfare economic theory*

Viewed from the perspective of conventional welfare economic theory one might be inclined to argue – and many have already done so – that to undertake any project involving a cost that exceeds its expected benefit would involve a suboptimal use of society's scarce resources – see for example House of Lords (2006), paras 62, 63 and Jones-Lee and Aven (2010). From this it would seem to follow as a matter of logical necessity that since costs can naturally be viewed as being "disproportionate" to benefits whenever they exceed the latter, then the stipulation that a safety improvement should be undertaken provided its costs are not grossly disproportionate to expected benefits is effectively to require that safety expenditure should be undertaken to a point well in excess of that which would be justified if the aim were to ensure an efficient use of scarce resources.

But of course this conclusion will be justified only if costs and benefits are appropriately defined and estimated (Jones-Lee and Aven, 2010). While in principle the costs of implementing a safety improvement are relatively straightforward to specify, defining and estimating the benefits of a safety improvement is altogether more problematic from both a conceptual and empirical point of view.

In previous chapters we have seen how the risk-reduction benefit is assessed. To combine the cost and benefits the approach of willingness-to-pay (WTP) is often used. The idea is that the benefits of a safety improvement should be defined in such a way as to reflect the preferences – and more particularly the *strength* of preference – of those members of society who will be affected by the safety improvement concerned (Jones-Lee and Aven, 2010). According to the WTP approach to the valuation of safety, the aim is to determine the amounts that affected individuals would be willing to pay for the (typically small) reductions in the risk of death or injury afforded by a particular safety improvement, given that the individuals concerned have been fully informed about the nature of the risk reductions. These amounts are then aggregated (possibly with distributional weights applied to reflect considerations of equity and fairness) across the affected group to arrive at an overall monetary value for the safety improvement concerned.

In order to standardise values of safety derived under this approach, the concept of the prevention of a "statistical" fatality or injury is employed (Jones-Lee and Aven, 2009). Thus, suppose that a group of 100 000 people enjoy a safety improvement that reduces the probability of premature death during a forthcoming period by, on average, one in 100 000 for each individual in the group. While the safety improvement could turn out to prevent no deaths, or one death (in fact, the most likely outcome) or two deaths (with a lower probability) and so on, the arithmetic mean (statistical expectation) of the number of deaths prevented is precisely one and the safety improvement is therefore described as involving the prevention of one "statistical" fatality.

Now suppose that individuals within this group are, on average, each willing to pay €x for the one in 100 000 reduction in the probability of death afforded by the safety improvement. Aggregate willingness to pay will then be given by €x times 100 000. This figure is naturally referred to as the WTP-based "value of preventing one statistical fatality" (VPF). An alternative term often used is the "value of statistical life" (VSL). Thus, if on average, the members of the population were willing to pay €15 per year to reduce their risks of death to this extent, the VPF (or VSL) would be €1.5m.

The VPF is closely linked to the ICAF. If the computed ICAF for a specific risk-reducing measure is larger than the VPF, i.e. ICAF > VPF, then this measure is not justified, and vice versa, if ICAF < VPF, the measure is justified. In a company context the WTP reflects the decision-maker's willingness to pay.

In order to avoid possible confusion it is very important to appreciate that, as defined above, the VPF is *not* a "value or (price) of life" in the sense of a sum that any individual would accept in compensation for the certainty of his or her own death – for most of us no sum, however large, would suffice for this purpose so that in this sense life is literally priceless. Rather, the VPF is in fact aggregate willingness to pay for typically *very small* reductions in individual risk of death (which, realistically, is what most safety improvements actually offer at the individual level).

For situations in which a proposed project can be expected to increase, rather than reduce, risk for some section of the public, the ethical precepts underpinning social cost–benefit analysis clearly entail that the cost of the increased risk should be defined on the basis of the minimum amount that those adversely affected would be willing to accept as compensation for the deterioration in their personal safety. By now there is ample empirical evidence that "willingness to accept" (WTA)-based costs of risk will exceed their WTP-based counterpart values of safety by a factor of between three and five (Jones-Lee and Aven, 2009).

Consider the LNG case and the possible measure of removing the LNG plant (at an early stage of the development). A quick analysis of this measure would, however, quickly reveal that that the expected number of saved lives (ICAF) would be small compared to the costs and gross disproportion would clearly be documented when comparing to standard values of ICAF. An alternative approach would be to derive WTP/WTA figures directly from people on this measure (relocate the plant). The implementation of the measure is judged worthwhile provided the total WTP/WTA exceeds the cost of the safety measure.

Such an approach assumes that people have the ability and the adequate information needed to judge uncertainties and declare their preferences accurately. However, this presupposition can be questioned (Jones-Lee and Aven, 2009):

Non-Revelation of Preferences: if we ask respondents about the WTP/WTA for implementing the measure, many would say that such a measure should be paid for by the authorities as they have the responsibility of providing a safe environment for all people in the community. Through the tax system people have already paid for a safe environment. Many people would have stated extreme compensation numbers for accepting a new risk, as it would favour their view.

Limited Information: The WTP approach presupposes well-informed and carefully thought-out preferences, and given the rather high degree of uncertainty associated with the plant operation many would argue that the survey would be a number crunching exercise without a rationale.

Hence the best we can do seems to be to use indices such as ICAF, and the risk assessment provides input to this use. The VSL is a reference for deciding which measures to implement. However, as discussed above, this approach is based on expected values and does not adequately reflect uncertainties. Jones-Lee and Aven (2010) discuss this issue from a WTP perspective. They argue that the gross disproportion principle can be justified due to the uncertainties, but also because the societal concerns are not fully reflected in WTP-based values (as normally defined). Societal concerns relate to the wider adverse impact, on society as a whole, that might result from the occurrence of a large-scale accident or hazardous event and which one could reasonably expect might be largely ignored by the typical individual operating in the narrowly focused rôle of "private citizen". It is referred to in HSE (2001) which defines societal concerns as follows:

Societal concerns [are] the risks or threats from hazards which impact on society and which, if realised, could have adverse repercussions for the institutions responsible for putting in place the provisions and arrangements for protecting people, e.g. Parliament or the government of the day. This type of concern is often associated with hazards that give rise to risks which, were they to materialise, could provoke a socio-political response, e.g. risk of events causing widespread or large scale detriment or the occurrence of multiple fatalities in a single event. Typical examples relate to nuclear power generation, railway travel, or genetic modification of organisms.

Again we refer to the approach summarised in Figure 1.4 which is developed to meet these challenges (this approach uses VSL as a reference but also gives weight to uncertainties and other concerns).

Finally in this section we show how the ALARP principle can be used in a terrorist risk management context.

7.2.2 *Applying the ALARP principle to terrorist risk management*

ALARP is a commonly used framework for managing risk due to non-intelligent threats, but terrorism introduces difficult issues, both technically and socially. Guikema and Aven (2010) discuss these issues and their implications for risk management. They argue that despite the challenges posed by adaptive threats, ALARP is still a useful and well-defined framework for risk management for adaptive threats, provided that the costs and benefits are defined in a broad enough manner and that the displacement of risk to other types of attacks is explicitly accounted for. An approach for verifying ALARP inspired by Figure 1.4 is presented in the following (Guikema and Aven, 2010):

The main ideas are summarised in these points:

1. Perform an initial crude qualitative analysis:
   (a) Perform a crude qualitative analysis of the local benefits and burdens of the risk-reducing measure.
   (b) Perform a crude qualitative assessment of the potential for additional risk to be imposed on others due to attacker substitution in response to the risk-reducing measure.
   (c) Perform a qualitative assessment of the loss of civil liberties associated with the risk-reduction measure.

If the local costs are not judged to be large relative to the local risk-reduction benefits, if there are no risks imposed on others that are judged to be above the broadly acceptable threshold for individuals, and if there is judged not to be a significant loss of personal liberties, implement the risk-reduction measure. Gross disproportion has not been demonstrated.

2. If the costs are considered large, quantify:
   (a) The conditional risk reduction where the analysis is conditioned on the attack occurring. Perform an economic analysis as indicated above, computing for example E[NPV] or ICAF given an attack.
   (b) The additional risk imposed on others due to attacker substitution in response to risk-reduction measures.

If E[NPV] > 0, or ICAF is low for the case of unconditional assessments (provided that such calculations have been carried out),
   or
if E[NPV | attack] is judged to be high or ICAF | attack is judged to be low and the attack probability is judged to be non-negligible, implement the measure provided that there is judged to be no significant loss of personal liberties. Gross disproportion has not been demonstrated.

3. If these criteria are not met or if a conditional approach has been used in the second step, assess uncertainty factors and other issues of relevance not covered by the previous analyses. A checklist is used for this purpose. Aspects that could be covered by this list are:
   (a) Is there considerable uncertainty (related to phenomena, consequences, conditions, and background knowledge that the attack likelihoods are based on) and will the measure reduce these uncertainties?
   (b) Does the measure significantly increase manageability?
   (c) Is the measure contributing to obtaining a more robust solution?
   (d) Is the measure based on best available technology (BAT)?
   (e) Are there unsolved problem areas: personnel safety-related and/or work environment-related?
   (f) Are there possible areas where there is conflict between these two aspects?

If the risk-reducing measure scores high on these factors (many yes answers), gross disproportion has not been demonstrated provided that the civil liberties assessment of step 4 below is passed.

4. If gross disproportion has not be demonstrated by steps 1–3, assess whether or not the burden from any loss of civil liberties imposed by the risk-reduction measure is grossly disproportionate to the risk reduction achieved by the measure. This is an inherently political judgement and cannot be based on purely technical risk considerations. If potential loss of civil liberties is not grossly disproportionate to the risk reduction, gross disproportion of overall costs and benefits has not been demonstrated.
5. If gross disproportion has not been demonstrated, implement the measure.

This approach imbeds traditional, quantitative ALARP assessment based on cost–benefit analysis within a larger qualitative framework that aims to address the many difficult-to-quantify aspects of risk reduction for intentional threats. Some of these other factors are inherently political and well beyond the realm of quantitative risk analysts' expertise. This points to the need for broad stakeholder engagement in this process of ALARP assessment (Guikema and Aven, 2010).

This framework can be based on the (A,C,U) risk perspective. All probabilities used are knowledge-based probabilities, and expectation is with respect to these probabilities. Chances (frequentist probabilities) cannot be meaningfully defined in this case (Aven and Renn, 2009b, see also comments on this issue in the coming section). Hence the (A,C,P_f) perspectives cannot be used as a platform for this example.

7.3 The role of the cautionary and precautionary principles

The analysis in Chapters 5 and 6 has clearly shown the need for dealing with uncertainties – in estimates and in the background knowledge that the probabilities are conditional on. We cannot just refer to the probabilities computed, and base our decision on these numbers. We need to give weight to the uncertainties, in other words give weight to the cautionary principle. As mentioned in Section 1.2, the precautionary principle is considered a special case of the cautionary principle, as it is applicable in cases of *scientific uncertainties* about the possible consequences of the activity being considered (Aven, 2006).

In this section we will discuss how these principles are affected by the findings in Chapters 5 and 6. To this end we will make use of a risk-uncertainty classification system presented in Stirling and Gee (2002); see Figure 7.1. Based on this system, Stirling and Gee (2002) provide an interesting discussion of the precautionary principle, and the uncertainty dimension in particular, by seeing the uncertainty dimension (referred to as incertitude) in relation to the strengths and weaknesses of risk assessments, as well as to the fundamental dimensions of incertitude (risk, uncertainty, ambiguity and ignorance). The precautionary principle applies when we have poor knowledge about the likelihoods, and the outcomes are poorly defined, i.e. category IV in Figure 7.1 (refer also to Stirling, 1998, 2007). Precaution is also relevant to some degree for categories II and III.

In the following discussion we refer to the four categories in Figure 7.1 as I, II, III and IV – we would like to avoid the reference to risk and uncertainty in categories I and II as such a terminology is in conflict with the common

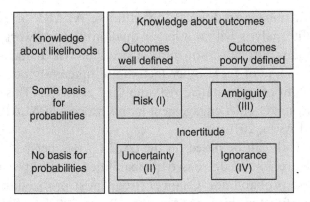

Figure 7.1   A classification system for incertitude (Stirling and Gee, 2002).

interpretation of risk (Holton, 2004; Aven, 2003). The risk–uncertainty distinction in Figure 7.1 is inspired by or based on the ideas of Knight (1921) which was mentioned in Section 2.4: under risk the probability distribution of the performance measures can be assigned objectively, whereas under uncertainty these probabilities must be assigned or estimated on a subjective basis. However, adopting this terminology, we cannot speak about risk in most practical applications, as objective probability distributions cannot be determined. For example, *terrorism risk* as a term would make no sense in this conceptual framework. And the precautionary principle would not be a part of risk management as this principle extends far beyond the narrow term *risk* used in the Stirling and Gee (2002) classification.

Stirling and Gee (2002) adopt a modification of the Knightian framework by restricting risk to situations with well-defined outcomes and some basis for the probabilities, i.e. not a requirement of objective probabilities. Nonetheless, such a convention would be in conflict with risk assessment being a tool to express uncertainties, and the reference to the four dimensions of incertitude (risk, uncertainty, ambiguity and ignorance) is, therefore, not used in the following analysis.

We interpret Figure 7.1 as shown in Figure 7.2 and explained in the following.

7.3.1 The objective of risk assessment is uncertainty description

The knowledge dimension being poor (strong) means that the basis for assigning the knowledge-based probabilities is poor (strong), i.e. the background knowledge K is poor (strong). For category III, only the probabilities related to A are relevant as the outcome space for C is poorly defined.

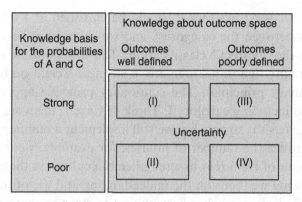

Figure 7.2   Implementation of the Stirling and Gee (2002) classification structure under the risk perspective (A,C,U) when the objective of the risk assessment is uncertainty descriptions.

The outcome space dimension relates to scenarios and not only the final outcome. For example, if the activity is exposure to electromagnetic radiation, we may easily define some overall injury/fatality categories, but it would be difficult to define a complete set of effects and scenarios as the underlying phenomena are not fully understood. The consequences of electromagnetic radiation are an example of category III uncertainties. The outcome space is poorly defined, but the probabilities (related to A – i.e. exposures) could have a strong basis. The discussion in Chapter 6 is relevant for situations of all categories, but only the (A,C,U) risk approach is relevant for categories III and IV as, if the outcome space is poorly defined, adequate chance distributions would be difficult to establish.

7.3.2 The objective of the risk assessment is to accurately estimate the risk (probabilities)

When the objective of the risk assessment is to accurately estimate the risk (probabilities), the knowledge dimension relates to the accuracy of the probability estimates. For category I (and III with respect to A) the estimates are relatively accurate (small epistemic uncertainties), whereas for category II (and IV for specific As) the estimates are subject to large epistemic uncertainties relative to the true underlying probabilities. Figure 7.2 is still relevant with "knowledge basis for the probabilities of A and C" replaced by "knowledge about the probabilities of A and C". The discussion in Chapter 5 is relevant only for situations of the categories I and II.

The case analyses in Chapters 5 and 6 are best described as belonging to the categories I and II. The LNG case is the one with highest uncertainties, but it can be discussed whether it should be classified as I or II. There are no sharp lines between the categories, and we should not be too concerned about finding the "correct" classification. The key point is that increased level of uncertainty (ignorance) requires a stronger weight on the cautionary and precautionary principles. The cautionary principle applies in all four categories, also including category I. Think of Case 1, the working accidents. The knowledge basis is strong, but we still implement a number of cautionary measures to reduce the number of injuries; for example, we may use a considerable amount of resources to strengthen the culture in the organisation.

In the following we focus on the understanding and use of the precautionary principle. This principle has been subject to strong debate and its link to risk assessment is not straightforward. Precaution means that actions are taken in situations of scientific uncertainties about the consequences of an activity. Hence, it is essential for regulators and the industry to define the concept of scientific uncertainties. The many definitions of the precautionary principle provide different suggestions for how to understand this concept, and the topic has been given due attention in the literature. It is, however, difficult to conclude when looking at all these definitions and reading the literature.

Here are some common interpretations of "scientific uncertainties" (Aven, 2010g):

1. Large uncertainties exist in outcomes relative to the expected values.
2. There is a poor knowledge basis for the assigned probabilities.
3. There are large uncertainties about frequentist probabilities (chances) p.
4. It is difficult to specify a set of possible consequences (state space).
5. There is a lack of understanding of how the consequences (outcomes) are influenced by underlying factors. It is difficult to establish an accurate prediction model (a cause-effect relationship).

In addition, it is possible to define combinations of these interpretations, as in the Stirling and Gee (2002) case where 2 and 4 are used as the criterion.

The essential feature of the precautionary principle when looking at the many definitions of the principle is that it should apply when the consequences of an activity could be serious but we do not fully understand what could happen. Thus, there must be a potential for surprises. These ideas provide the basis for the coming discussion when we make judgements about the above interpretations being adequate. The discussion is based on Aven (2010g).

Adopting this criterion we can quickly conclude that large uncertainties in outcomes relative to expected values are not sufficient for applying the precautionary principle. Most real-life situations are characterised by variation leading to large differences between the potential outcomes and the expected values, but the underlying phenomena and processes may still be well-understood. Consider the working accident, case 1. The risk assessment has computed a probability distribution for the number of serious injuries for a period of one million manhours (Figure 5.1) and a related expected value. The distribution has a relatively large variance, reflecting the fact that the number of injuries could be large. In this case there are large uncertainties about the actual number of injuries, and consequently also in relation to the expected value. But we would not classify these uncertainties as scientific as the potential for surprises is considered minor. The situation is characterised as one in category I, using the Stirling and Gee (2002) classification (see Figures 7.1 and 7.2).

Let us consider the other interpretations, and let us think about the LNG case. Suppose the knowledge basis for assigning the leakage probabilities is considered relatively poor, in the sense that few relevant data are available and the adopted model of the underlying phenomena is rather crude. Are the uncertainties then scientific uncertainties?

In this case the understanding of the underlying phenomena and processes is strong in the sense that a cause–effect relationship can be established: we understand how the consequences (outcomes) are influenced by underlying factors. In theory we can construct a prediction model for the occurrence of leakages and the consequences, which would give accurate predictions when we know the input parameters. To formalise this, let Z be the output quantity of interest and X a vector of model input quantities (parameters). Then a model G can be defined such that $G(X)$ produces accurate predictions of Z. Assuming that such a prediction model can be established, would we classify the uncertainties as scientific uncertainties?

What matters should not be the background knowledge used for assigning the probabilities, but the total knowledge basis about the phenomena and processes considered. If an accurate prediction model can be established, the knowledge basis is strong and we cannot refer to scientific uncertainties. There is not a potential for surprises. Establishing such a model also means that it is possible to define the state space for the outcomes. On the other hand, if such a model cannot be established, it is reasonable to refer to the uncertainties as scientific uncertainties.

The Stirling and Gee (2002) classification links scientific uncertainties mainly to category IV (refer to Stirling, 1998): no basis for the probabilities, and outcomes poorly defined (difficult to define a state space for the outcomes). Using IV as the basis for comparison, a stronger definition is obtained than interpretation 5 (an accurate prediction model cannot be established) since poorly defined outcomes mean that an accurate prediction model cannot be established. The definitions are not equivalent as we can construct cases where interpretation 5 applies, but not category IV. For instance, if an accurate prediction model cannot be established, the uncertainties may be in category II: the outcomes are well-defined but there is a poor knowledge basis for A and C. Specific examples include floods under climate change, and shareholder value (refer to Stirling *et al.*, 2006).

The definition of frequentist probabilities (chances) p means a modelling of the analysed phenomena, as p is a fraction of successes in a constructed population of similar units to the one (those) studied. Hence, if such probabilities (chances) have been defined, the uncertainties cannot be scientific uncertainties according to interpretation 5. The introduction of such probabilities (chances) means that we have some knowledge about the phenomena studied. There is not a potential for surprises.

Think of the following situation: you are offered a game where you do not know the probabilities (chances) of the lottery machine producing the prize/ loss money. In this case frequentist probabilities (chances) p can be defined and hence the uncertainties are not scientific uncertainties according to interpretation 5 (but according to criteria 1–3 they are). In the Stirling and Gee (2002) classification, the uncertainties are of type II. The state space is well-defined: the real line $(-\infty, \infty)$, and there is no basis for the probabilities. If the type of prize/loss is not known, the uncertainties become scientific uncertainties, according both to interpretations 4 and 5 and to the Stirling and Gee (2002) classification. A chance distribution cannot be established and the outcomes are poorly defined.

As another example, let us look at the frequentist probability (chance) of a terrorist attack. Such a probability (chance) does not exist, as a large (infinite) population of similar situations cannot be meaningfully defined (Aven and Renn, 2009b). Knowledge-based probabilities can, however, be specified. Based on the available knowledge at a particular point in time the analyst (expert) may assign a knowledge-based probability of an attack equal to 0.1 (say), meaning that he/she considers the uncertainty (degree of belief) to be comparable to randomly drawing one specific ball out of an urn comprising ten balls. When making such an assignment it is essential to be precise on what the basis for the assignment (the background knowledge) is. For example, we may assume that the possible attackers do not have access to information about the uncertainty assessments and risk assignments, or the measures taken to follow up these assessments and assignments. If such access is available, this may influence the possible attackers and requires

Table 7.1 *Main findings of the analysis of the five interpretations 1–5 of scientific uncertainties.*

Interpretation	Sufficient for situation to be classified as scientific uncertainties
1. Large uncertainties in outcomes relative to the expected values	No
2. A poor knowledge basis for the assigned probabilities	No
3. Large uncertainties about relative frequency-interpreted probabilities (chances) p	No
4. It is difficult to specify a set of possible consequences (state space)	Yes (4 implies 5)
5. It is difficult to establish an accurate prediction model	Yes

new and updated uncertainty assessments (Aven, 2007a, refer also to Lindley 2006, pp. 75–76).

The knowledge basis for such knowledge-based probabilities would often be poor. We may for example have little information available of when and how an attack could occur. Hence, criterion 2 applies. Yet the consequences of an attack could be rather accurately predicted. The uncertainties are therefore not scientific uncertainties according to criteria 4 and 5. In the Stirling and Gee (2002) classification, the uncertainties are again of type II. It is of course possible to think of potential attacks involving phenomena where the consequences are not easily predicted and the outcome space is not well-defined; then the uncertainties become scientific uncertainties, according both to interpretations 4 and 5 and to the Stirling and Gee (2002) classification. However, the most common situations are "surprising attacks", not "surprising consequences".

In Table 7.1 we have summarised the main findings of the above analysis.

7.3.3 An alternative classification system

An alternative to the Stirling and Gee (2002) classification system is presented in Figure 7.3, based on the idea that scientific uncertainty is related to the difficulty of establishing a prediction model for the consequences. The classification system presented relates the scientific uncertainties to the *cause–effect relationship*, which is a basis for many of the existing definitions of the precautionary principle. By introducing different categories of the strength of the cause–effect relationship, we are able to distinguish between different levels of scientific uncertainties. In this set-up, Z denotes the quantity of

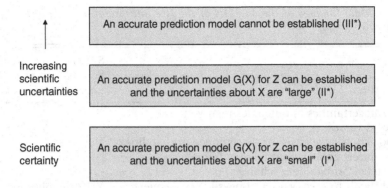

Increasing
scientific
uncertainties

Scientific
certainty

> An accurate prediction model cannot be established (III*)
>
> An accurate prediction model G(X) for Z can be established and the uncertainties about X are "large" (II*)
>
> An accurate prediction model G(X) for Z can be established and the uncertainties about X are "small" (I*)

Figure 7.3   Different categories of uncertainties, reflecting various degrees of scientific uncertainties. Z is the output and $X = (X_1, X_2 \ldots)$ are the underlying factors influencing Z.

interest and $X = (X_1, X_2 \ldots)$ are the underlying factors influencing Z, and G is the model linking X and Z. Note that scientific certainty in this sense does not mean that Z can be predicted with accuracy when not conditioned on X. Unconditionally, the consequences Z are uncertain, and this uncertainty is defined by the uncertainties of the factors X. To make a judgement about the uncertainties of X being small, different approaches can be used, for example:

- an accurate prediction model for X can be established
- a chance distribution of X can be established showing the variation of X if the situation considered can be repeated over and over again.

To explain this set-up and these approaches in more detail we will consider the safety design case 3, alternative II, the parallel system comprising two components.

Let Z be the relative downtime of the system in a test interval of length τ. To analysis Z we introduce the model $G(X) = \max\{\tau - \max\{X_1, X_2\}, 0\}/\tau$, where X_1 represents the time to the failure of the component, i.e. G is equal to the relative time from when the second unit fails until τ. Thus we put $Z = G(X)$. Now suppose that we have a lot of relevant data about the performance of the components. These data provide the basis for determining chance distributions F_1 and F_2 for X_1 and X_2, respectively. The components are judged to be independent.

We conclude that the situation is characterised by scientific certainty. An accurate prediction model can be established. There are uncertainties about X_1 and X_2, but the variations in lifetimes are known.

Now suppose that the amount of relevant data (information) is rather limited. Chance distributions F_1 and F_2 are established but their justification is not so strong. Parameters θ_1 and θ_2 are introduced, and we write $F(x \mid \theta_1)$ and $F(x \mid \theta_2)$ to show that the distributions depend on the parameters θ_1 and θ_2. The available knowledge provides wide uncertainty bands (confidence intervals or credibility intervals if the Bayesian perspective is adopted) for the parameters. Under these conditions, the uncertainties about X are much larger than in the previous example, and the situation is classified as category II*.

Next suppose that it is difficult to establish a specific family of chance distributions for X. Then the uncertainties about X are even larger but still categorised as "large". The scientific uncertainties are also larger, but are not categorised as level III* as the model G(X) provides insights on how X influences Z.

Now let us look at a slightly more complex situation (Aven and Vinnem, 2007): the consequences of a blowout in Norwegian oil and gas production in the Barents Sea. Consider the consequences of an oil spill on fish species, and let Z denote the recovery time for the population of concern, with Z being infinity if the population is not recovered. Then there is scientific certainty according to I* if we can establish a function (model) G such that Z equals G(X) with high confidence, where $X = (X_1, X_2 \ldots)$ are some underlying factors influencing Z, and the uncertainties about X are "small". Such factors could relate to the possible occurrence of a blowout, the amount and distribution of the oil spilled on the sea surface, the mechanisms of dispersion and degradation of oil components, and the exposure and effect on the fish species. For values of X, we can use G to predict the consequences Z.

In this case there are considerable uncertainties about some components of X and the model G is disputed among experts. Many biologists conclude that there is a lack of fundamental understanding of the underlying phenomena concerning the effect on the fish species – an accurate prediction model G cannot be established. However, others argue that such a model can be constructed. Potential surprises can be ignored. The fact that many experts in the field conclude that an accurate prediction model cannot be constructed leads to a classification of the situation in category III*. The highest institution in the Norwegian Church referred to the precautionary principle when they requested the Norwegian government not to start year-round petroleum operations in the Barents Sea some years ago. The Norwegian government gave strong weight to the principle in their decision-making, and not all fields were opened for year-round operations.

A classification system, such as the Stirling and Gee (2002) framework (Figures 7.1 and 7.2) and the above system (I*–III*), characterises and structures different types of uncertainties. In this way the system may improve our understanding of the concept of "scientific uncertainties", and next, guide us in determining what should be the key features defining "scientific uncertainties" in applications. It could also serve as a communication tool between different stakeholders, in particular between analysts, people with policy and legal backgrounds, and the decision-makers. For these systems to work, key concepts, such as probability, need to be precisely defined. Such precision is in place for both classification systems studied above. The classification system suggested explicitly incorporates the basic building blocks of risk assessment: models and probabilities. It distinguishes between aleatory uncertainties represented by probability models and chances, and epistemic uncertainties expressed by knowledge-based probabilities. In this way, this classification system represents a rise in the level of detail and precision compared to the Stirling and Gee (2002) system. In our view the suggested system not only provides insights important for the proper invocation of the precautionary principle, but also practical guidance for what to look for when determining what scientific uncertainties mean. The classification is not based on precise limits for when the situation should be categorised as I*, II* and III*. Judgements have to be made in each case.

7.4 Risk communication

The form and content of the risk assessments, and in particular the "associated uncertainties", influence the risk communication, which could be defined as

an interactive process of exchange of information and opinion among individuals, groups and institutions. It involves multiple messages about the nature of risk and other messages, not strictly about risk, that express concerns, opinions or reactions to risk messages or to legal and institutional arrangements for risk management.
(US National Research Council 1989, p. 21)

In short, we could say that risk communication means the exchange of risk-related knowledge and information between the stakeholders.

The ultimate goal of risk communication is to assist stakeholders and the public at large in understanding the rationale of risk-informed decisions, and to arrive at a balanced judgment that reflects the factual evidence about the matter at hand in relation to their own interests and values (Aven and Renn, 2011). In other words, good practices in risk communication are meant to help all affected parties to make informed choices about matters of concern to them. The purpose of risk communication should not be seen as an attempt to convince people, such as the consumers of a risk-bearing product, that the communicator (e.g., a government agency that has issued advice concerning the product) has done the right thing. It is rather the purpose of risk communication to provide people with all the insights they need in order to make decisions or judgements that reflect the best available knowledge and their own preferences (Aven and Renn, 2011).

This way of looking at risk communication is supported by the (A,C,U) type of risk perspective. Uncertainties, the knowledge and lack of knowledge, need to be revealed and communicated. It can also be supported by the (A,C, P_f) perspectives when the aim is accurate risk estimation, but this perspective often leads to a use where uncertainties are camouflaged or hidden, and the risk assessment is used for risk-based decision-making. Again we could think about the LNG case. By reference to the calculated probabilities, the operator concluded in the actual situation that the risk was acceptable and this was communicated to all other parties, including the neighbours. Uncertainties were not reported. In fact the term "uncertainty" is not referred to at all in the main risk analyses report. The basic philosophy seems to be that the risk is very low and the neighbours would acknowledge this if informed by the experts used by the operator. The situation reminds us about the time when experts informed people about the risk of nuclear accidents. The message was that the probability of a serious accident (meltdown) is so small that it does not represent a problem. The calculated probability of a serious incident is $1/1\,000\,000$ (say), it is so low so we need not be concerned. More knowledge would lead lay people to draw the same conclusion.

However, the judgement about acceptability is a value judgement, and it is based on evaluations of both potential consequence and probabilities/ uncertainties. What is important here is not just the likelihood. Rejection of nuclear power plants can be justified based on the fact that potential consequences are extremely large and that even a small probability makes the risk unacceptable.

It is a common conception, in particular among analysts, that it is difficult to communicate probabilities and uncertainties: lay people do not understand these concepts, nor do managers and politicians. This view can, however, be challenged. Our perception is that people can appreciate these concepts if they are properly introduced. Unfortunately that is often not the case. If the analysts preparing the risk and uncertainty picture do not have a solid understanding of the key concepts, they will fail in transforming any message to a broader audience. The scientific platform of risk assessment as described in this book could help in communicating the results, as precision is required on what are the uncertain quantities and what the tools are to express the uncertainties. In the actual LNG case, the presentation and communication

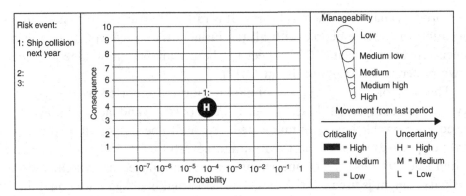

Figure 7.4   Bubble diagram for event ship collision next year (based on Abrahamsen and Aven, 2010).

of the risk results lacked a foundation. If you introduce a probability to express uncertainties, you must explain what that means. The analysts (experts) have assigned a probability equal to 0.2 (say), but what does this mean? The assessments produce formulae and numbers but hardly any comments and reflections on the tool used to describe the risk.

Many analysts seem to think that information provided to managers and politicians needs to be very simple and not include discussions of uncertainties. This is, however, a misconception. Managers and politicians are able to relate to and deal with uncertainties and risk, yes, these tasks are largely what their jobs are all about – to make decisions under uncertainty and risks. Managers are usually well-equipped people who will quickly understand what is at stake and what the key issues are if the professionals can do their job. The problem is rather that the analysts have not been able to report the uncertainties and present them in an adequate way. A lack of competence on the scientific platform of risk assessment among analysts and other professionals often results in inadequate uncertainty assessments.

To inform the decision-makers and other stakeholders about risk, different visualising tools could be used. An example of such a tool is bubble diagrams (Abrahamsen and Aven, 2011; Abrahamsen *et al.*, 2010). In a traditional bubble diagram, risk is shown through three dimensions: (1) consequence, (2) probability and (3) manageability, but it is also possible to include (4) uncertainties that extend beyond the assigned probabilities/expected values as shown in the example in Figure 7.4. The diagram reflects the four dimensions by showing the probability on the x-axis (normally using a logarithmic scale), the consequences on the y-axis, the manageability is visualised through the bubble size, and the uncertainty dimension using a letter reflecting the assigned uncertainty category. The criticality of the risks is determined based on an assessment of all dimensions, and is represented by a colour. The classification of the risks in the bubble diagram is just a snapshot of the situation and is continuously updated. The various dimensions are defined as follows:

Probability

The probability included in the diagram is a knowledge-based probability of an event A conditional on a background knowledge K.

Consequence

The "consequence" dimension is to be interpreted as the expected negative impact given the occurrence of the event A.

Manageability

The easiness with which risk can be reduced and desirable outcomes ensured. The "easiness" relates to the organisation's capability to reduce risk and obtain desirable outcomes seen in relation to other concerns, in particular cost. We say that the manageability is high if it is considered feasible to implement measures over time which can reduce risk and give increased confidence in obtaining desirable outcomes. Similarly we understand a low manageability.

Uncertainty

Uncertainty reflects the expected values' predictability of the real outcomes. High uncertainty in the bubble diagram may for example express that the assigned expected number of fatalities can give a poor prediction of the actual number of fatalities given the occurrence of A.

In the example three uncertainty categories are used, defined as follows (Flage and Aven, 2009):

Low uncertainty

All of the following conditions are met:

- the phenomena involved are well understood; the models used are known to give predictions with sufficient accuracy
- the assumptions made are seen as very reasonable
- much reliable data are available
- there is broad agreement among experts
- low variation in populations (low stochastic uncertainty)

High uncertainty

One or more of the following conditions are met:

- the phenomena involved are not well understood; models are non-existent or known/believed to give poor predictions
- the assumptions made represent strong simplifications
- data are not available, or are unreliable
- there is lack of agreement/consensus among experts
- high variation in populations (high stochastic uncertainty)

Medium uncertainty

Conditions between those characterising high and low uncertainty, e.g.:

- the phenomena involved are well understood, but the models used are considered simple/crude
- some reliable data are available

Note that the degree of uncertainty must be seen in relation to the effect/influence the uncertainty has on the predicted consequences. For example, a high degree of uncertainty combined with high effect/influence on the predicted values will lead to the conclusion that the uncertainty factor is important. However, if the degree of uncertainty is high but the predicted values are relatively insensitive to changes in the uncertain quantities, then the uncertainty classified in the diagram could be low or medium.

The bubble diagram is closely related to a risk matrix. In the bubble diagram there will be a unique classification of the risk since attention is given to expected consequences. For "ship collision next year" the risk will be classified in the bubble diagram as the point $(P(A), E[C \mid A])$. This way of classifying risks can also be adopted for a risk matrix, but it is also common to use consequence categories in risk matrices. For example, if a ship collision occurs we may consider the consequence categories C_1 (0 fatalities), C_2 (1–5 fatalities), C_3 (6–20 fatalities) and C_4 (more than 20 fatalities).

We may start the criticality classification by first ranking the risks according to the three standard dimensions consequence, probability and manageability. Then we may adjust these up or down in case the uncertainties are considered high or low.

The risk-related information visualised in bubble diagrams could be a useful communication tool in safety management. The diagrams summarise important features of the knowledge and lack of knowledge available and are continuously updated. The accuracy of the method is in line with the precision of the assessment tool.

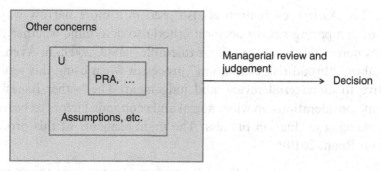

Figure 7.5   An illustration of the three categories of input to the managerial review and judgement. "PRA, ..." refers to the formal probabilistic risk assessment and other probability-based methods (for example cost–benefit analyses founded on expected net present value calculations), the U refers to the uncertainties not fully captured by the assessments.

Many other similar diagrams can be established, for example showing the expected number of lives saved and expected costs, in addition to uncertainties; see Abrahamsen *et al.* (2010). This diagram could facilitate communication between analysts and other stakeholders of safety measures' cost-effectiveness.

7.5 The content and purpose of managerial review and judgement

The results of risk assessments are used to support decision-making as discussed in Section 1.1; see Figures 1.1–1.3. The decision emerges from a managerial review and judgement phase during which the management/ decision-makers consider

(i) the formal results of risk assessments and other probability based assessments
(ii) the premises, assumptions and limitations of these assessments
(iii) other issues not captured by the assessments (e.g. strategic issues).

Figure 7.5 illustrates these three categories of input to the managerial review and judgement.

The weight given to the uncertainties, i.e. the cautionary and precautionary principles, is often decisive for the decision that is made.

Concerning item (ii), consideration should be given as to which decision alternatives have been analysed and the fact that the results of the analyses represent judgements (expert judgements), etc. (refer to the list in Section 1.1).

In a risk management/governance context, the managerial review and judgement phase is closely linked to the risk evaluation process (Aven and

Renn, 2010). As risk evaluation is also used in a more narrow way as "a process of comparing risk to decision criteria to determine whether the risk and/or its magnitude is acceptable or tolerable" (ISO, 2009a,b; Aven, 2003), we talk about "broad risk evaluation" processes when using this term as an alternative to managerial review and judgement. The "other issues" cover important considerations on wider social and economic factors to be included in the balancing evaluation process. The main elements of this process are (Aven and Renn, 2010):

- Summarising the results of the risk assessment process in terms of a risk description, also including characterisations of uncertainties.
- Deliberation over these results in consideration of wider social and economic factors (e.g. benefits, societal needs, quality of life factors, sustainability, distribution of risks and benefits, social mobilisation and conflict potential), legal requirements and policy imperatives).
- Weighing of pros and cons and trading-off of different (sometimes competing or even conflicting) preferences, interests and values.

The broad risk evaluation assesses "high-level" value-based issues, such as the choice of technology, societal needs requiring a given risk agent to be present and the potential for substitution as well as for compensation, and reaches beyond the risk itself and into the realm of policy-making and societal balancing of risks and benefits.

While risk assessment deals with knowledge claims (around what are the causes and what are the effects), evaluation deals with *value claims* (around what is good, acceptable and tolerable). This concept has been criticised as being too simplistic and inadequate for complex risk problems. Indeed, the distinction between non-tolerable, tolerable and acceptable is quite simple but it reflects the actual need for a judgment in many situations. The LNG case is an illustrating example. This final closure on the risk allows for only three alternatives: either to do nothing, to ban the risk or to initiate risk-modifying actions. There is no other alternative at this point. This important judgement should be made as transparent as possible to all interested individuals and parties and the organisations responsible for this judgement need to have the skills, the assets, the background knowledge and the sensitivity with respect to the corresponding values and socio-cultural preferences to arrive at an informed, balanced and fair judgement (Aven and Renn, 2010).

As stressed throughout this book decision-making under risk and uncertainties should be risk-informed, not risk-based, i.e. the managerial review and judgement has a role to play. Nonetheless, risk-based approaches are common in practice and the literature includes a vast number of theories and methods, as well as applications based on this thinking. Risk assessment consultants and the formal decision-making on the regulation of risk remain relatively unaffected by this recognition (Stirling, 1998).

There are many reasons for this, but a main factor is obviously lack of understanding of the fundamentals of risk assessments and risk management. The limitations of the risk assessments need to be taken into account as well as other concerns, see Figure 7.5.

Closely related to this lack of understanding of the fundamentals of risk management and risk assessments there is an issue about the risk managers' and the risk analysts' mind set concerning the use of risk assessments in the decision-making process (Aven 2010i):

Too many risk managers want to be absolved of the responsibility for having to make decisions – they want a "risk-based" approach in which their actions are dictated by risk assessments. Then, if their actions turn out to be wrong (poor), they can claim absolution on the basis that "We did what the numbers told us to do. If the numbers were wrong, it's the analysts' fault." Risk analysts, for their part, too often lack sufficient analytic humility and fall into the trap of trying to give answers with far greater certainty than can be justified. This is the real issue in the risk-based versus risk-informed dichotomy.

7.5.1 How the risk perspective affects the managerial review and judgement

In the rest of this section we will look more closely into the managerial review and judgement and the related components (i)–(iii) introduced above, based on the analysis of the scientific criteria in Chapters 5 and 6. The key issue is how the risk assessment perspectives, with their strengths and weaknesses, influence the managerial review and judegment.

Let us first study the (A,C,P_f) risk perspective where the aim of the risk assessment is accurate risk estimation, i.e. of P_f. Suppose that a substantial amount of relevant data is available and accurate estimates of the parameters have been computed. Think of the working accident data of Case 1 and a decision problem related to possible implementation of risk-reducing measures. In this case (i) is covered by the assessments in Chapter 5, and to a large extent (ii) is also covered. In addition, considerations have to be given to costs, as well as other concerns such as political aspects. To fully appreciate the third item (iii) we need to understand the detailed decision situation considered and its framing. As an illustration, consider again the Case 1 example. Two "other concerns" in this case include (Aven *et al.*, 2010):

Political signals: In the Norwegian context the Storting (the Norwegian parliament) and Cabinet give guidance to the Norwegian Petroleum Safety Authority and the industry in white papers. Here some issues are frequently repeated and highlighted. Firstly the petroleum industry is described as a leading industry which continuously invests in knowledge and improvement by learning from best practice. "A levelling or decline in Health Safety and Environment (HSE) performance is not in line with such objectives" the papers state. Secondly, the introduction of the "Zero-Philosophy" is seen as a milestone regarding attitude and behaviour in the industry. Thirdly the obligation of implementing international rules and regulations is stated. Such guidance contributes to priorities and focus areas for the regulating authorities, both in revising and developing the regulations, the inspection and instructing or guiding the industry.

The "Nordic Model" of a healthy working life (Kettunen, 1998) presupposes an *active participation* of the workers in job design, running operation and risk assessment. It is also embedded in the Working Life Act and safety regulation. A major concern from the authorities in the development of new regulations is to have a continuous focus on the role of workers and their unions as an educated and motivated resource to improve safety work. Arenas have been established in the industry where groups of representatives from the various interested parties discuss and review important safety issues, and in particular the results from risk assessments and other expert judgments.

Thus for the decision-maker to decide on implementation or not of a specific measure, it is not sufficient just to look at the risk and cost assessments with evaluations of their premises, assumptions and limitations. The other concerns may strongly influence the conclusion. The situation considered may be one where the industry has experienced some events (near misses) which had a potential for severe consequences and the safety climate is strongly affected by these events. Then this climate could mean a stronger willingness to use resources on safety than is normal. Many other such factors (e.g. the economic climate) are relevant for the managerial review and judgement.

Even in Case 1 where the amount of relevant data was extensive, the results are strongly dependent on the assumptions made. We remember for example from Chapter 5 how the serious injury rate was influenced by the assumption of a trend. Seeing the risk assessment as a tool for informing the decision-maker it is essential that the risk assessments seek to provide broad risk characterisations reflecting different sets of assumptions. In view of the uncertainties of the risk estimates produced, this type of sensitivity analysis is required and management (decision-makers) should always ask for such analyses to be carried out for key assumptions. It is a common attitude among many analysts and experts that they should provide clear recommendations on what the decision-maker should do. However, the analysts and experts should appreciate their role in the decision-making process as informers, nothing more. The managerial review and judgement normally extend far beyond the context of the analysts and experts.

Now let us look at the situation where the aim of the risk assessment is uncertainty characterisations. What we have said above also applies to this situation. In many ways, it is easier to meet the requirements of sensitivity analysis showing the dependencies of key assumptions, as this perspective is based on uncertainty descriptions rather than telling the "truth" about risk. As there is no "truth" it is easier for the analysts and experts to see their roles as risk informers. For the LNG case, the actual execution of the risk assessments was based on a perspective where the risk assessments seek to present the "real" risk picture, although as we noticed from the analysis in Chapters 5 and 6, this is a misuse of the risk assessments. Such a perspective cannot be justified. The neighbours and many independent experts did not find the risk characterisation sufficiently informative to support the decision-making on the location and design of the LNG plant. Sensitivity analyses were lacking, as well as reflections on uncertainties (a sensitivity analysis is not the same as an uncertainty analysis, see discussion in Aven, 2010a). The risk figures produced are based on a number of critical assumptions, but these assumptions are not integrated in the risk characterisation presented nor communicated by the operator (Aven, 2009c).

The managerial review and judgement was reduced to (i) – a more or less mechanical transformation of the results of the assessments to conclusions about risk acceptability. Emphasis on (ii) would have provided a broader basis for the decision-making, as shown by the analysis in Chapter 6 where risk perspectives are applied, revealing and describing uncertainties. Following the (A,C,U) perspective, where uncertainty factors are presented and discussed, in addition to probability-based indices and sensitivity analyses, the managerial review and judgement would cover all three items (i) –(iii). The risk description would give more weight to the uncertainties. This does not necessarily mean a different decision, but it could.

From this it is obvious that the choice of risk perspective is important for the decision-making and risk management, as was also discussed in Section 1.2. It may serve the interests of the operator to present a risk picture that concludes that the real risk is small and acceptable. Focusing on uncertainties would easily create an image of the operator being unknowledgeable. Such a thinking is common although there is broad recognition among risk experts that typical risk assessments provide a narrow picture on risk and it is important to see beyond the assessments when managing risk; see Chapter 1. There are many reasons for the often narrow perspective adopted, but a main factor is certainly the risk assessment tool in itself – the scientific quality of

risk assessment as discussed in previous chapters. However, equally important is the role of the assessment in the decision-making context. It simply
serves the interests of many actors to maintain narrow risk descriptions as
these will provide a stronger support for their desired decision alternatives.
We have mentioned the LNG case, but the literature includes many others.
Perhaps the most common type of examples relate to the nuclear industry. In
the early stage of the development of this industry, the experts' description of
risk was narrow and uncertainties were not properly reported, as was mentioned also in Section 7.4.

Also a broad risk perspective highlighting uncertainties can be misused. If
intolerable risk is sought, a possible strategy is to put more emphasis on the
uncertainties than justified. A constant and strong focus on the uncertainties
from many actors would easily lead to a higher risk perception among
managers and politicians, as risk is dependent on the uncertainties. To avoid
such misuse it is essential that the risk and uncertainty assessments are carried
out according to some sound standards as for example outlined in this book.
Failures to do this could have serious implications for the decision-making
processes, as we have discussed in previous sections and chapters, and further
discuss in the following.

In the LNG case, the neighbours of the plant as well as many experts
have constantly stressed the uncertainties. And some politicians have been
concerned about the risk level. The issue is to what extent weight is given to
the cautionary principle. The precautionary principle is not an issue as the
uncertainties are not so much about scientific uncertainties. However, the
argumentation and rationale supporting the judgements have not been
strong enough to cause the politicians to reverse their decision on accepting
the operation of the plant. Perhaps this is not so surprising given the fact
that the national safety agency had no objections to the operator's risk
assessment approach. There are also strong economic incentives for not
interrupting the planned activities. The role of the safety agency is critical.
Often the agencies have a focus more on whether a risk assessment has
been carried out than on its quality and whether it provides meaningful
decision support.

Vinnem (2010) states that the approach taken by the agency may be
interpreted as lack of competence or lack of professional maturity. In the
case of the LNG plant the risk analysis was used to "prove" that it was safe
enough not to follow the US practice for safety zones for LNG plants.
However, such a practice is strongly criticised by others, including the safety
agency for the petroleum industry in Norway. In November 2007 this agency
issued a letter to the industry warning about the malpractice of using risk

analysis to demonstrate the acceptability of deviations from accepted practice and regulatory requirements (Vinnem, 2010).

7.5.2 Climate Change example

Now let us consider a case on a "higher societal level", the implications of global climate change. The international expert group Intergovernmental Panel on Climate Change (IPCC) has gone through considerable effort to articulate a common characterisation of climatic risks and uncertainties. Given the remaining uncertainties and the complexities of the causal relationships between greenhouse gases and climate change, it is then a question of values as to whether governments place their priorities on prevention or on mitigation (Keeney and McDaniels, 2001).

From the risk characterisation of the IPCC Fourth Assessment Report, Climate Change 2007, we give some examples of typical statements:

- Warming of the climate system is unequivocal, as is now evident from observations of increases in global average air and ocean temperatures, widespread melting of snow and ice, and rising global average sea level.
- Rising sea level is consistent with warming. Global average sea level has risen since 1961 at an average rate of 1.8 [1.3 to 2.3] mm/yr and since 1993 at 3.1 [2.4 to 3.8] mm/yr, with contributions from thermal expansion, melting glaciers and ice caps, and the polar ice sheets. Whether the faster rate for 1993 to 2003 reflects decadal variation or an increase in the longer-term trend is unclear.
- Observational evidence from all continents and most oceans shows that many natural systems are being affected by regional climate changes, particularly temperature increases.
- There is *medium confidence* that other effects of regional climate change on natural and human environments are emerging, although many are difficult to discern due to adaptation and non-climatic drivers.
- Global GHG (greenhouse gases) emissions due to human activities have grown since pre-industrial times, with an increase of 70% between 1970 and 2004.
- Global atmospheric concentrations of CO_2, methane (CH_4) and nitrous oxide (N_2O) have increased markedly as a result of human activities since 1750 and now far exceed pre-industrial values determined from ice cores spanning many thousands of years.
- There is *very high confidence* that the net effect of human activities since 1750 has been one of warming.

- Most of the observed increase in globally-averaged temperatures since the mid-20th century is *very likely* due to the observed increase in anthropogenic GHG concentrations. It is *likely* there has been significant anthropogenic warming over the past 50 years averaged over each continent (except Antarctica).
- Human influences have:
  - *very likely* contributed to sea level rise during the latter half of the 20th century
  - *likely* contributed to changes in wind patterns, affecting extra-tropical storm tracks and temperature patterns
  - *likely* increased temperatures of extreme hot nights, cold nights and cold days
  - *more likely than not* increased risk of heat waves, area affected by drought since the 1970s and frequency of heavy precipitation events.
- Continued GHG emissions at or above current rates would cause further warming and induce many changes in the global climate system during the 21st century that would *very likely* be larger than those observed during the 20th century.

These statements represent the panel's risk description or characterisation. There are large uncertainties and these are reflected by the likelihood judgements. This risk characterisation provides a basis for broad risk evaluations carried out by political parties and governments in different countries, as well as international organisations. Most parties judge the present level of GHG emissions to be intolerable, and formulate targets of x per cent cut from current levels by year y, to obtain a tolerable level (Aven and Renn, 2010).

7.5.3 Different decision settings

Decisions involving uncertainty and risk are made at different organisational levels and in a number of settings. Process plant managers encounter situations which force them to make decisions that will seriously affect production goals and accident risk in a conflicting manner. To make satisfactory decisions, they are dependent on decisions by senior management, e.g. in the form of policy statements, about priorities of accident risk versus production goals. Regulatory agencies can be seen to make decisions when imposing new requirements, e.g. to perform risk analysis and deal with risk in specified ways. It is obvious that the context and nature of the decision processes mentioned vary significantly. Often, decision-makers are constrained in a way which does not allow them to assess risk in detail. For example, in a crisis the time constraints do not allow for detailed risk assessments.

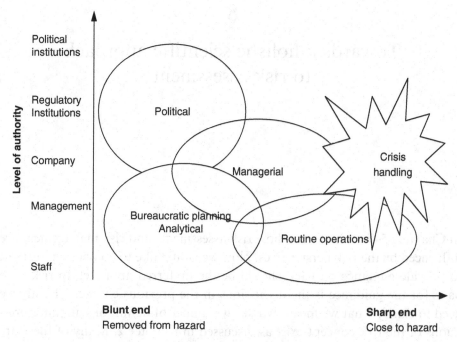

Figure 7.6   Classification of decision settings (Kørte *et al.*, 2002).

These constraints are closely related to the decision settings. Many classification systems for such settings have been developed. In Rosness (2009) (see also Kørte *et al.* (2002) and Aven (2003)), a system is presented based on a taxonomy of decision-makers. The classification is based on two dimensions: closeness to hazard and level of authority. See Figure 7.6. Decision settings typical for certain groups of actors are identified and the constraints for these are discussed. The implications these constraints have on decision-makers or actors with respect to risk analysis and management are considered and the necessity for interaction among actors in different decision settings is shown.

The form and content of the managerial review and judgement depend on the decision setting. The risk assessments as well as "other concerns" could vary considerably. As an example, contrast the LNG case and the climate change case. However, the overall principles and ideas are the same. The risk assessments provides decision support, and due considerations have to be given to the framing of the assessments, their assumptions and limitations, and the other concerns that are important for the decision-making. We have to acknowledge that there is no simple and mechanistic method or procedure for balancing different concerns.

8

Towards a holistic scientific approach to risk assessment

In Chapters 5–7 we have seen how risk assessments and risk management are influenced by the risk perspectives. Now we would like to go one step further, to provide guidance on what should be the preferred approach to risk. The basis for the guidance is the discussions in the previous chapters. Firstly we need to clarify what we mean by risk. A number of definitions and interpretations of the risk concept exist as discussed in Chapter 2. Many of these are probability-based. Below (Section 8.1) we present and discuss a structure for characterising the definitions, which is founded on a clear distinction between (Aven, 2010f)

(a) risk as a concept based on events, consequences and uncertainties;
(b) risk as a modelled, quantitative concept; and
(c) risk descriptions.

The discussion leads to an approach for conceptualising and assessing risk, which is based on risk defined by (a), i.e. is founded on the (A,C,U) risk perspective, and the probability-based definitions of risk can be viewed as model parameters and/or risk descriptions. The approach provides clear guidance on how to think when conceptualising and assessing risk in practice.

Next in this chapter (Section 8.2) we present and discuss a general model-based framework for risk assessments. Starting from an industry guide to quantitative uncertainty analysis and management, clarifications and simplifications are made to ensure consistency with the (A,C,U) risk perspective. Some simple examples are included to motivate and explain the basic ideas of the framework.

In risk assessments, probability is the common tool used to describe the epistemic uncertainties about unknown quantities. However, the purely probability-based approaches to risk and uncertainty analysis can be challenged as we have discussed throughout this book. A key point is that the support of the probabilities is not reflected by the numbers produced. This concern has sparked a number of investigations in the field of uncertainty representation and analysis, which has led to the development of several alternative approaches, including possibility theory and evidence theory. These theories and methods represent strong research areas and in the last section of this chapter (Section 8.3) we question to what extent the raised challenges of the probability-based methods can be solved by these approaches.

8.1 What is risk? A structure for conceptualising and describing risk

Risk is a fundamental concept for most scientific disciplines, but no consensus exists on how to define and interpret risk. Some definitions are based on probabilities, some on expected values, and others on uncertainty. Some consider risk as subjective and epistemic, dependent on the available knowledge, whereas others grant risk an ontological status independent of the assessors. The situation is chaotic and leads to poor communication. We are also afraid that it hampers effective risk management as well as the development of the risk field, as many of these definitions and interpretations lack proper scientific support and justification.

Of course, business needs a different set of risk methods, procedures and models from, for example, medicine and engineering. But there is no reason why these areas should have completely different perspectives on how to think when approaching risk and uncertainty, when the basic challenge is the same – to conceptualise that the future performance of a system or an activity could lead to outcomes different from those expected, desired, planned, or not in line with stated objectives.

Think of an activity in the future, say the operation of an offshore installation for oil and gas processing. We all agree that there is some risk associated with this operation. For example, fires and explosions could occur leading to fatalities, oil spills, economic loss, etc. But it is not straightforward to explain what we mean by this risk if we require a precise definition and would like to use the concept in scientific studies. Risk analysts would introduce a set-up which directly or indirectly defines how risk is understood and assessed; refer to Chapters 5 and 6. The set-up would typically be probability-based, with probabilities interpreted either as relative frequencies or as subjective probabilities. All such set-ups can be challenged as not being able to reflect risk in a proper way. Important risk aspects could be camouflaged or hidden by the set-up. Discussions of the set-up are therefore important, not only from a theoretical point of view but also from a practical risk management perspective.

We have identified several definitions of risk that can be used as an overall, common definition. They all belong to the category (a). Many attempts have been made to establish a unified risk perspective, but none of these have obtained broad acceptance in practice. There could be many reasons for this. Firstly, the scientific work on risk may not have reached a sufficiently mature level for establishing such a definition. The exploring phase is not completed. Secondly, the scientific literature has a focus on the generation of new ideas and suggestions, and on a critique of other contributions. By its nature, it is hard to obtain broad consensus on scientific issues in general and risk definitions in particular. And thirdly, the standardisation organisations have not been able to produce sufficient broad and precise definitions which could be accepted by the scientific expertise on risk.

Consider for example the latest definition from the International Standardisation Organisation (ISO, 2009a,b): risk is the effect of uncertainty on objectives. What does this mean? Risk has to do with uncertainty, but is it the *effect* of uncertainty? And risk is related to objectives, but what if objectives are not defined? Then we have no risk? Asking experts on risk, there is no doubt that this definition would lead to numerous different interpretations. The definition is not sufficiently precise, and one may certainly also question its rationale as indicated.

In Chapter 2 we presented and discussed a set of common definitions of risk, including (the numbers 1–8 are the same as those used in Chapter 2)

0. Risk equals the expected loss (Verma and Verter, 2007; Willis, 2007).
1. Risk is a measure of the probability and severity of adverse effects (Lowrance, 1976).
2. Risk is the combination of probability and extent of consequences (Ale, 2002).
3. Risk is equal to the triplet (s_i, p_i, c_i), where s_i is the ith scenario, p_i is the probability of that scenario, and c_i is the consequence of the ith scenario, $i = 1, 2, \ldots N$ (Kaplan and Garrick, 1981).
4. Risk refers to uncertainty of outcome, of actions and events (Cabinet Office, 2002).
5. Risk is a situation or event where something of human value (including humans themselves) is at stake and where the outcome is uncertain (Rosa, 1998, 2003).
6. Risk is an uncertain consequence of an event or an activity with respect to something that humans value (IRGC, 2005).
7. Risk is equal to the two-dimensional combination of events/consequences and associated uncertainties (Aven, 2007a, 2010e).

8. Risk is uncertainty about and severity of the consequences (or outcomes) of an activity with respect to something that humans value (Aven and Renn, 2009a).

For the measures that are based on probabilities and expected values, we may generate two versions, one where the probabilities are interpreted as relative frequencies (and the expected values as averages), and one where the probabilities are subjective (knowledge-based) probabilities (and the expected value is interpreted as the centre of gravity of the probability distribution). We write definitions x_f and x_s, respectively, to separate the two categories, $x = 0, 1, 2, 3$. Consider as an example category 0, risk defined as the expected loss. According to definition 0_f, risk is understood as the average loss when considering an infinite number of similar situations, whereas 0_s means that risk is the centre of gravity of the subjective probability distribution of the loss. Following the suggested structure for characterising the various risk definitions we have to place these definitions in one of the categories (a), (b) (c), defined above.

The result is that definition 0_f is in category (b) and 0_s is in category (c), as risk in the former case is based on the model of an infinite number of similar situations and risk in the latter case is a way for the assessor to describe or characterise risk. The expected loss E_s when using subjective probabilities is a risk index based on the background knowledge (K) of the assessor. A similar analysis is carried out for the other eight definitions. The result is shown in Table 8.1.

We refer to Chapter 2 for a discussion of these and other risk definitions.

Table 8.1 *Categorisation of the nine risk definitions according to the structure (a)–(c)*

Risk definition	Category
0_f	b
0_s	c
1_f	b
1_s	c
2_f	b
2_s	c
3_f	b
3_s	c
4	a
5	a
6	a
7	a
8	a

If relative frequency-interpreted probabilities P_f constitute the basis (definitions 0_f, 1_f, 2_f, and 3_f) risk is a modelled, quantitative concept (category b) and we may formalise the definitions by writing

$$Risk = (A, C, P_f),$$

where A represents the events (initiating events, scenarios) and C the consequences of A, as in Section 2.2.

If, on the other hand, subjective (knowledge-based) probabilities constitute the basis (definitions 0_s, 1_s, 2_s, and 3_s), the definitions must be viewed as risk descriptions as they express the analysts' (experts') degree of belief concerning A and C. Also the background knowledge K that the probabilities are based on should be considered a part of the risk description.

If we search for a widespread agreement on one definition of risk we have to look among the categories (a). The others have to be excluded as they are based on either a model or an assignment of uncertainty using the tool, subjective (knowledge-based) probability. Risk should also exist as a concept without modelling and subjective probability assignments. We face risk when we drive a car or run a business, also when probabilities are not introduced. For risk assessment we need the probabilities, but not as a general concept of risk. In this way we obtain a sharp distinction between risk as a concept and risk descriptions (assessments).

As discussed in Section 2.5, definition 4 (which basically says that "risk = uncertainty") cannot be used as it fails to include the consequence dimension. Hence we are led to two candidates among the a-definitions: the (A,C,U) definitions (7–8) and the (A,C) definitions (4–6). The latter group means that the common risk terminology has to be revamped (refer to discussion in Section 2.5) and we therefore prefer to use the (A,C,U) definition.

Risk is thus defined. The next stage would then be to specify how to describe risk. We seek a general structure and we cannot base it on the use of frequentist probabilities (chances) as these cannot be meaningfully defined in all cases. However, knowledge-based probabilities can always be defined, and they are introduced as the recommended tool for describing the uncertainties.

This leads to a risk description as was first noted in Section 2.8:

$$Risk\ description = (A, C, U, P, K),$$

that is, risk is described by events A and consequences C, subjective (knowledge-based) probabilities P, uncertainties U not captured by P, and K the background knowledge that U and P are based on. The U component may for example be a qualitative assessment of uncertainty factors (assumptions that the probabilities are based). A subjective probability $P(A) = P(A \mid K)$ is interpreted as a knowledge-based probability with reference to

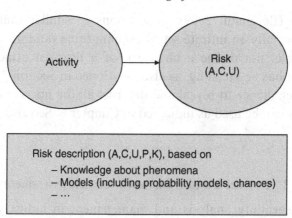

Figure 8.1   The main elements of the recommended risk approach.

an uncertainty standard expressing the assessor's uncertainty about the occurrence of the event A given the background knowledge K. Following this interpretation the assessor compares his/her uncertainty (degree of belief) about the occurrence of the event A with the standard of drawing at random a favourable ball from an urn that contains $P(A) \cdot 100\%$ favourable balls (Lindley, 2000).

However, also in this setting we may establish relative frequencies, but they are referred to as chances and not probabilities. A chance is the limit of a frequency of similar (formally exchangeable) random events. More generally we introduce probability models with unknown parameters. A chance is an example of such a parameter. By the Bayesian updating machinery, knowledge about the parameters is described first by the prior distribution, then updated to produce the posterior distribution to reflect observations. Finally, this distribution is used to generate the predictive distribution of the events A and consequences C. These predictive distributions then incorporate the variation reflected by the probability model (and the chances) and the epistemic uncertainties about the true value of the parameters. The main features of the thinking are shown in Figure 8.1. Models will be further discussed in the coming section.

Note that chances and probability models are tools used to describe risk. They are not identified as risk per se. This is in contrast to the "Risk $= (A,C,P_f)$" types of approaches, including the probability of frequency approach (see Section 2.5), where the relative frequency-interpreted probabilities (chances) P_f always need to be defined. They constitute the foundation of the approach. In the (A,C,U) types of approaches, chances are only defined when exchangeable sequences can be justified. Chances need some sort of model stability (Bergman, 2009): populations of similar units need to be constructed (formally an infinite set of exchangeable random variables). We will, for example, not define a chance p of a terrorist attack (Aven and Renn, 2009b); it has no meaning, as also mentioned in Section 7.3.

It may be a challenge to reveal and describe all the uncertainties. Qualitative approaches can be used as indicated in Chapter 6. See also the discussion in Section 8.3.

8.2 A model-based framework for risk assessments

A guide to uncertainty analysis and management in industry has recently been issued (de Rocquigny *et al.*, 2008). The guide is written by a project group of the European Safety, Reliability and Data Association (ESReDA). The book project was motivated by the fact that no authoritative standard exists for how to analyse and quantify uncertainty. The guide presents a numbers of practical cases, all based on the same uncertainty analysis framework; see Figure 8.2. As uncertainty is a main component of risk as defined in the previous section, this framework for uncertainty assessment is highly relevant to risk assessments. The discussion in this section is to a large extent based on Aven (2009c, 2010b).

The key variables of interest are denoted Z (which could be a vector). To assess Z a model $G(X,d)$ is introduced which links a set of input variables X and some fixed quantities d to Z (also X and d could be vectors). To describe the uncertainties, probabilistic and non-probabilistic methods (for

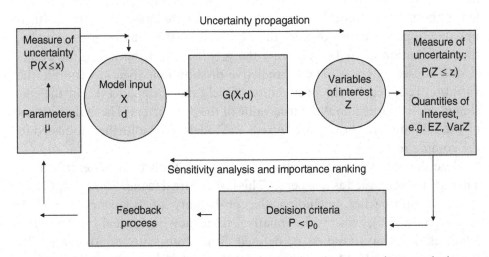

Figure 8.2   The overall framework adopted by the uncertainty analysis guide (de Rocquigny *et al.*, 2008).

example possibility theory and evidence theory, see Section 8.3) can be used. A common approach is to use a parametric probability distribution (where μ is the parameter) to establish a probability distribution for X. Using the model G, an uncertainty description is obtained for Z. The tool used for this purpose could be an analytical approach or Monte Carlo simulation. Some quantities of interest, for example expected values and variances, are specified and computed from the measure of uncertainty derived, i.e. the probability distribution of Z. These quantities provide input to a decision process, which could be based on some decision criteria expressing for example that a probability should not exceed a specified level. Sensitivity analysis provides insights about how the input quantities affect the output quantities, and importance ranking identifies what factors, subsystems etc. are most important based on some defined criteria, for example the contribution to the variance. The result of the analysis may lead to some action (feedback process), for example that there is a need for design changes to meet the criteria. The actions need to be seen in relation to the goals of the analysis which usually fall into the following categories (de Rocquigny *et al.*, 2008):

Understand: To understand the influence or rank the importance of uncertainties, and thereby to guide any additional measurement, modelling or research and development efforts.

Accredit: To give credit to a model or a method of measurement, *i.e.* to reach an acceptable quality level for its use. This may involve calibrating sensors, estimating the parameters of the model inputs, simplifying the system model physics or structure, fixing some model inputs, and finally validating according to a context-dependent level.

Select: To compare relative performance and optimise the choice of maintenance policy, operation or design of the system.

Comply: To demonstrate compliance of the system with an explicit criterion or regulatory threshold.

Most analysts and researchers would probably consider this framework a logical and useful structure for performing uncertainty analysis in practice. There is not much that is controversial or problematic about the framework described at this overall level. However, when we go into the details, the meaning and use of the different concepts are not so straightforward as we will see from the coming analysis.

Risk analysis may be considered more restricted than uncertainty analysis as risk analysis focuses on future events, whereas uncertainty analysis is concerned with uncertain quantities, whether they relate to the future or not. However, the frameworks and tools used for analysing risk are to a large extent general and in most cases they are applicable also for "non-future" type of situations. Anyhow, uncertainty is the key concept to be addressed and we need to clarify:

(i) What are the uncertain quantities?
(ii) Who is uncertain?
(iii) How should we represent the uncertainties?

In the framework illustrated by Figure 8.1 the uncertain quantities are X and Z, but in practice it may not be straightforward to choose the appropriate X and Z as we have seen from the discussions in Chapters 5 and 6. Are X and Z observable quantities like time to failures and costs, or parameters of probability models? Uncertainty about the quantities X and Z raises the question: who is uncertain? Is it the decision-maker, the analyst or some experts used in the assessment? We will argue that the uncertainty is normally that of the analyst. Experts and others can produce input to the analyst but the analyst has the ownership of the final distributions and quantities of interest. Of course, in some cases the aim of the analysis is simply to report the knowledge expressed by some experts, but as the analysts are responsible for how to elicit this knowledge and analyse it, care should be shown in presenting the results as independent of the analysts. Being precise on the ownership is essential for obtaining a clear understanding of the framework and how to communicate its results. For a further discussion of this issue, see Aven and Guikema (2010).

To express the uncertainties an adequate representation is required, and probability is the natural choice as it meets some basic requirements for such a representation (Bedford and Cooke, 2001, p. 20):

- *Axioms*: Specifying the formal properties of the uncertainty representation.
- *Interpretations*: Connecting the primitive terms in the axioms with observable phenomena.
- *Measurement procedures*: Providing, together with supplementary assumptions, practical methods for interpreting the axiom system.

Many types of uncertainty representations exist, but many fail when it comes to interpretation. We should reject a representation which has no clear interpretation. It is not sufficient to say that a measure expresses for example a degree of belief. We need to know what it means that the measure is 0.2 instead of 0.4.

The present analysis has a focus on the use of probability to measure uncertainty, although the de Rocquigny *et al.* (2008) framework allows for both probabilistic and "non-probabilistic" representations of uncertainty. We refer to the discussion in Section 8.3.

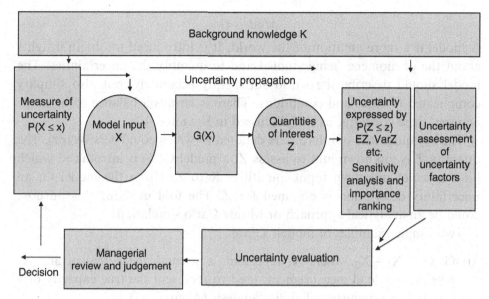

Figure 8.3 Structure of a modified framework (based on Aven 2009c, 2010b).

8.2.1 A modified framework

It is possible to simplify ideas and clarify key concepts in the de Rocquigny *et al.* (2008) framework when restricting attention to probabilities as a measure of uncertainty. See Figure 8.3. The basic features of the modified framework are described in the following.

The quantities X and Z are well-defined

The quantities X and Z must express states of the "world", i.e. quantities of a physical reality or nature, that is unknown at the time of the analysis but will, if the system being analysed is actually implemented, acquire some value in the future, and possibly become known (Aven, 2003). The quantities X and Z must have some true, objective values. No ambiguity can be present. In our view, uncertainty assessment of quantities for which true and precisely defined values do not exist, cannot be a basis for a scientific risk assessment. This is a key assumption of the framework, and supported by for example Bedford and Cooke (2001).

If chances (relative frequency-interpreted probabilities) are introduced they must be considered unknown properties of the world and be treated as X and Z in the framework. If chances are introduced, meaningful interpretations must be possible.

Models G

A model is a representation of the world. It is introduced to obtain insights about the phenomena being studied and to quantify the uncertainties. The model should describe the world sufficiently accurately, but also simplify complicated features and conditions. There is always a balance to be made between these concerns, as was discussed in Section 6.4.1.

The key quantities of interest are denoted Z (which could be a vector). The quantity Z is unknown and to assess Z, a model G(X) is introduced which links a set of unknown input quantities X to Z. Using the model G, an uncertainty description is obtained for Z. The tool used for this purpose could be an analytical approach or Monte Carlo simulation.

Two simple examples of models G are:

(i) $G(X) = X_1 - X_2$, where X_1 represents a strength measurement and X_2 represents a load measurement, used to represent the true capacity of a system in a structural reliability analysis (Aven, 2003).
(ii) The exponential distribution $G(t| \lambda)$, used to represent the lifetime distribution $F(t)$ of a mass produced unit (i.e. the proportion of units with lifetime equal to or less than t). The parameter λ is interpreted as the inverse of the average lifetime in the infinite population of the units. In this example $X = \lambda$ and $Z = F(t)$.

Many other models are presented in Chapters 5 and 6.

For situations as in example (ii), the standard procedures for Bayesian analysis and Bayesian updating (Singpurwalla, 2006; Bedford and Cooke, 2001; Aven, 2003) are implemented as shown in Chapter 6

How to deal with model uncertainty is discussed in Section 6.4.1.

Probabilities are knowledge-based probabilities with reference to an uncertainty standard

All probabilities P introduced in the framework are knowledge-based (subjective) probabilities with reference to an uncertainty standard expressing the assessor's uncertainty about unknown quantities X and Z. Following this interpretation the assessor compares his/her uncertainty about the occurrence of the event A with the standard of drawing at random a favourable ball from an urn that contains $P(A) \cdot 100\%$ favourable balls (Lindley, 2000).

The probabilities $P(Z \leq z)$ etc. cannot be relative frequency-interpreted probabilities (chances) P_f as such probabilities are in fact not somebody's measure of uncertainty, but a way of expressing variation within a real or thought-constructed infinite (or very large) population of similar units to those studied. As an example, let P_f be the chance that a technical component fails during a specific period of time. This "probability" is understood as the fraction of components that fail in this period when considering an infinitely large population of similar components (we assume that such a population can be defined). In general P_f is an unknown population fraction, and as noted above has to be treated as X and Z in the framework. Consequently, $P(Z \leq z)$ etc. of the framework need to be interpreted as knowledge-based (subjective) probabilities.

These probabilities express *epistemic uncertainties*. The variation in the populations of similar units to the one studied, that for example generates the true value of P_f, is referred to as *aleatory (stochastic) uncertainty*.

But a knowledge-based (subjective) probability can be given different interpretations. Among economists and decision analysts, and the earlier probability theorists, a subjective probability is linked to betting. According to this perspective, the probability of the event A, $P(A)$, equals the amount of money that the assigner would be willing to put on the table if he/she would receive a single unit of payment in the case that the event A were to occur, and nothing otherwise. The opposite bet must also hold, i.e. the assessor must be willing to pay the amount $1 - P(A)$ if a single unit of payment would be given in return in the case that A were not to occur, and nothing otherwise. In other words, the probability of an event is the price at which the person assigning the probability is neutral between buying and selling a ticket that is worth one unit of payment if the event occurs, and worthless if not (Singpurwalla, 2006).

We argue that such an interpretation should not be used in a quantitative risk assessment context, as it extends beyond the realm of uncertainty assessments – it reflects the assessor's attitude to money and the gambling situation which means that analysis (evidence) is mixed with values. The scientific basis for risk assessment is based on the idea that professional analysts describe risk separated from how we (the assessor, the decision-maker or other stakeholders) value the consequences and the risk.

Assessments of uncertainty factors

Different probability-based measures are used to describe risk, such as the expected value, the variance and quantiles. But a full risk description needs to see beyond these P measures as has been discussed throughout this book, see for example Section 2.2. All probabilities are conditional on a background knowledge K, which includes assumptions and suppositions, and in particular the model G. This background knowledge is an integral part of the results of the analysis and all probabilities need to be considered in relation to K. The framework requires a separate identification and assessment of potential uncertainty factors hidden in K.

A qualitative assessment of these factors can be carried out as shown in Section 2.8 and in Chapters 5 and 6, addressing both sensitivity and uncertainty. A factor is given a high score according to this assessment if the risk indices are sensitive to changes in the factor, and the factor is subject to considerable uncertainties. To assess the degree of uncertainty the criteria mentioned in Section 7.4 can be used.

In the uncertainty evaluation, a broad uncertainty description is provided, covering probabilities and related background knowledge, as well as the results of sensitivity analyses. The evaluation provides input to a broader managerial review and judgement (corresponding to the feedback process in the framework of de Rocquigny *et al.* (2008)), which concludes on the implications of the analysis and balances different concerns. The result is for example an acceptance of the uncertainties related to an activity, the need for design changes, the choice of an alternative etc.

The decision-making context

In the de Rocquigny *et al.* (2008) framework, the quantities of interest are compared to relevant decision criteria, such as requirements of the form $P < p_0$, where p_0 is a fixed number. These requirements could for example express that the unreliability level of a piece of equipment should not exceed a specified level.

Now, given that P is a knowledge-based probability, can we justify basing our decision on a direct comparison of the form $P < p_0$? No, there is a need for a process that extends beyond the probabilistic analysis as was pointed out in Section 7.1. The probabilities are dependent on the background knowledge, the assumptions and suppositions made, including the model G. There is a need for a broader process (referred to as a managerial review and judgement) which sees the results of the assessment in a larger context, taking into account the limitations of the model, the difficulties in specifying probabilities for some quantities etc. It is a process extending beyond the domain of the uncertainty analysis. The sensitivity analyses constitute an important input to such a broad review and judgement process.

De Rocquigny *et al.* (2008) also acknowledge the need for broader evaluation processes – refer to the feedback process of Figure 8.2 – but stress that their framework focuses primarily on industrial situations in which there is enough modelling expertise, knowledge and/or data to support the use of quantitative modelling of risk and uncertainty, with probabilistic or mixed probabilistic/non-probabilistic tools. However, what is "enough" can always be discussed and even in cases with strong modelling expertise and much data

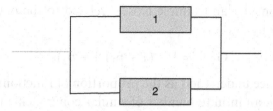

Figure 8.4   A parallel system comprising two components.

it is essential to avoid mechanical procedures for decision-making based on probabilistic (or other) criteria in isolation.

8.2.2 Examples of applications of the framework

Chapter 6 has in fact presented and discussed several applications of this uncertainty analysis framework. In Chapter 6 a distinction was made between risk defined through chances and risk defined through uncertainties. Both cases can be included in the set-up of the above framework. In the former case the chances P_f need to be treated as X and Z in the framework. Below we include another example, to summarise the basic ideas of the framework.

Reliability example

We consider a parallel system of two components, as shown in Figure 8.4. The state of the system is denoted Z, and is equal to 1 if the system is functioning and 0 otherwise. The system is functioning if at least one of the components is functioning. The parallel system defines a model G by

$$G(X) = 1 - (1 - X_1)(1 - X_2),$$

where X_i is the state of component i, $i = 1,2$, defined as 1 if component i is functioning and 0 otherwise.

Assuming independence between the component states we obtain

$$P(Z = 1 \mid K) = E(Z \mid K) = 1 - (1 - p_1)(1 - p_2),$$

where K is the background knowledge and $p_i = P(X_i = 1 \mid K)$, i.e. p_i equals the probability that component i is functioning.

This is a simple example of an application of the framework. It shows a standard reliability analysis where the probabilistic analysis adopts knowledge-based probabilities to express uncertainties about the states of the components and using the reliability block diagram (Figure 8.4) as a model.

Now let us consider an example closely related to the one above. Let us define the model by

$$G(p) = 1 - (1 - p_1)(1 - p_2),$$

where p_i is a chance understood as the proportion of functioning components when considering an infinite number of similar components to component i. The chances p_i are to be interpreted as the Xs in the framework. The quantity of interest in this case is another chance, the system reliability h, defined as the proportion of systems functioning when considering an infinite number of such systems. Hence $Z = h$ and by expressing knowledge-based probabilities P about the chances p_i we obtain an uncertainty distribution of Z. The analysis may be carried out using Monte Carlo simulation. The analysis is an example of a typical uncertainty analysis in a reliability and risk context, a probability of frequency analysis; refer to Sections 2.7 and 6.3.1.

A model may be introduced to express p_i, for example the exponential model, leading to

$$p_i = \exp\{-\lambda_i t\},$$

where λ_i is the failure rate and t is a fixed point in time. Based on a probability distribution for λ_i we obtain distributions for p_i and h. Following a standard Bayesian statistical analysis, we first specify a prior distribution for λ_i and then use the Bayesian updating procedure to obtain the posterior distribution when data become available.

If Z is the quantity of interest, we may first establish a distribution of (X_1, X_2) by expressions like

$$P(X_1 = 0 \text{ and } X_2 = 0) = \int (1 - \exp\{-\lambda_1 t\})(1 - \exp\{-\lambda_2 t\}) \, dH(\lambda),$$

where H is the probability distribution of $\lambda = (\lambda_1, \lambda_2)$. Then using the model $G(X) = 1 - (1 - X_1)(1 - X_2)$ we can compute $P(Z = 1)$. We thus run the framework twice, first for establishing the distribution of (X_1, X_2). In this case the exponential distribution is the model of the set-up and λ is the input quantity. In the second run, $G(X) = 1 - (1 - X_1)(1 - X_2)$ is the model and X is the input.

We end this section with some reflections important for the decision-making: what is gained by reducing the uncertainties before a decision is made? The question is related to the value of information.

8.2.3 *Value of information*

A decision is made at a specific point in time, but in many cases the decision-maker could defer the decision to gain more information and knowledge. The issue is then whether it is worth it. The answer would depend on many factors, including the cost of deriving the new information and knowledge, the costs of deferring the decision, and the consequences of making "the wrong" decisions. An assessment can be made of these factors to produce a basis for a decision on such a deferral. A risk assessment constitutes a main input in this process as it points to possible consequences of the deferral and provides judgements of the uncertainties. Then the various concerns need to be balanced and a decision made. In this sense the problem is similar to any decision problem under uncertainty and risk.

As a simple example, consider the reliability p (interpreted as the chance) of a production unit A. Suppose the chance p is subject to large uncertainties and a test programme is planned to reduce the uncertainties. To simplify, we assume that the testing will produce an accurate estimate of the "true" chance. The costs of the test programme are €1 million. The production income in the case where the reliability is p is specified by a function g(p). In the case that the testing reveals that $p < p_0$, a modification of the unit will be performed at a cost c and will increase p to p_0. We seek the value of the perfect information about p.

If p is such that $p \geq p_0$, no deferral is the best decision and the gain is g(p).

If p is such that $p < p_0$, the deferral means a gain $g(p_0) - c$ whereas the no deferral means a gain g(p), hence deferral is the best decision if $g(p_0) - c > g(p)$ and no deferral is the best decision in the opposite case.

Unfortunately, p is not known so we cannot use these statements as criteria. Instead it is common to use the expected value, leading to the following criterion (the value of perfect information):

The expected value in the case of no deferral equals E[g(p)]. The expected gain by perfect information is

$$E[I(p \geq p_0) \, g(p)] + E[I(p < p_0) \, \{(g(p_0) - c)I(g(p_0) - c > g(p)) + g(p) \, I(g(p_0) - c \leq g(p))\}]$$

where I(·) is the indicator function which equals 1 if the argument is true and 0 otherwise. This formula is based on the best decision for any p value. By comparing this term with E[g(p)], the decision-maker is guided on what to do.

This example fits into general decision analysis theory; see e.g. Lindley (1985) and Bedford and Cooke (2001). The basic pillars of this theory are the specification of utilities expressing the preferences of the decision-maker and subjective probabilities expressing her/his uncertainties. For a simple case like this the theory works nicely, but it is more difficult to implement in more complex situations as we have studied in this book. Specifying utilities is not straightforward. There will be a strong degree of arbitrariness in the choice of the utility function, and some decision-makers would also be reluctant to specify the utility function as it reduces their flexibility to weight different concerns in specific cases (Aven, 2010a).

For the cases considered in Chapter 3 we will have to approach the problem as indicated above, by running the decision process just as another risk decision problem, by assessing risk and other concerns, taking into account the limitations and constraints of these assessments and applying a managerial review and judgement before reaching a decision.

8.3 Probability and alternative approaches for representing (expressing) epistemic uncertainties

Using knowledge-based (subjective) probabilities to quantify uncertainties means that the analysts (experts) must express their degree of belief about unknown quantities using probability distributions. If it is known that a physical quantity has a value between 0 and 1, the assessor may for example specify a uniform distribution of [0,1] to express their uncertainties. The assessor then has assigned the same probability (1/2) for the quantity to be in the interval [0, 1/2] as [1/2, 1]. Following a probability-based approach such assignments are required. However, this perspective can be challenged; see e.g. Ferson and Ginzburg (1996) and de Rocquigny *et al.* (2008). It is argued that the assessments are based on unjustified assumptions. The information available for the probabilities does not provide a sufficiently strong basis for a specific probability assignment. In a risk analysis context, there are often many stakeholders and they may not be satisfied with a probability-based assessment providing subjective judgements made by one analysis group. In this section we will discuss this issue in more detail. Two main issues are addressed: (i) how to faithfully represent and express the knowledge available to best support the decision-making and (ii) how to best inform the decision-maker. The key references are Aven (2010c,d) and Aven and Zio (2011). We will relate this discussion to the reliability and validity criteria introduced in Section 3.3.

We face the issue of using probabilities to express lack of knowledge, i.e. epistemic uncertainties, in a risk assessment context. Many risk analysts consider probability to be the appropriate tool to represent such uncertainties, but there are different views. In recent years many books and papers have been published arguing that probability theory generates too precise results when the background knowledge of the probabilities is poor, and several alternative approaches have been presented (cf. Dubois, 2010; Aven and Zio, 2011):

(a) probability-bound analysis, combining probability analysis and interval analysis (Ferson and Ginzburg, 1996)
(b) imprecise probability, after Walley (1991) and the robust statistics area (Berger, 1994)
(c) random sets, in the two forms proposed by Dempster (1967) and Shafer (1976)
(d) possibility theory (Dubois and Prade, 1988; Dubois, 2006), which is formally a special case of the imprecise probability and random set theories.

As an illustration of the arguments put forward, consider an example similar to the one introduced by Ferson and Ginzburg (1996): a parameter θ_1 is an integer number between 1 and 5. How should the analyst describe this knowledge? Two different approaches are summarised:

Interval analysis: θ_1 is in the interval [1, 5], i.e. $\theta_1 \in \{1, 2, \ldots, 5\}$.
Probabilistic analysis: A uniform distribution is assumed for θ_1 over the set $\{1, 2, \ldots, 5\}$. Hence $P(\theta_1 = x) = 0.2$, for $x = 1, 2, \ldots, 5$.

We consider two such parameters: θ_1 and θ_2, where θ_2 is also an integer number between 1 and 5. There exist true underlying values of θ_1 and θ_2. The issue is how to describe the uncertainties of the product $\theta = \theta_1\theta_2$. The interval analysis produces an interval [1, 25] for θ and the probabilistic analysis produces a specific probability distribution $P(\theta = x)$ as illustrated by the distribution shown in Figure 8.5, assuming that θ_1 and θ_2 are independent.

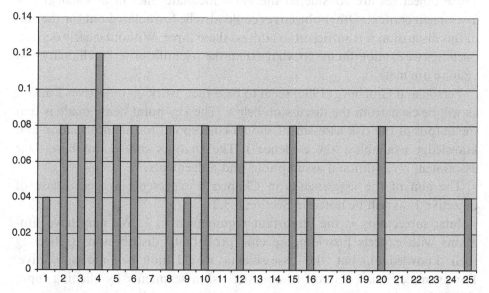

Figure 8.5 Probability distribution of $\theta = \theta_1\theta_2$ when θ_1 and θ_2 are independent.

Ferson and Ginzburg (1996) argue that the probabilistic distributions are incorrect, because they assume more information than was given in the original question. They write:

In this sense, they are the result of wishful thinking, rather than a careful analysis of what is actually known. This example illustrates what may be a widespread problem with applying classical probability theory in risk analyses where the relevant empirical information is sorely incomplete (as is usually the case).

(Ferson and Ginzburg, 1996)

They conclude by stating that probability theory provides the methods appropriate for assessing and propagating random variability, but not for assessing and propagating epistemic uncertainties and ignorance.

Such arguments are often being put forward and there is a need for a discussion of their strengths. To this end it is essential to clarify what the objective of the risk assessment is:

1. *to obtain an "inter-subjective" knowledge description of the unknown quantities*
2. *to obtain a faithful representation of the uncertainties reflecting the information and knowledge available, or*
3. *to obtain judgements about the unknown quantities from a qualified group of people (the analysts/experts).*

These objectives are considered the most adequate ones in a practical risk assessment context. Other objectives could also be formulated but for the aim of this discussion it is sufficient to address these three. Without clearly defined purposes we cannot discuss to what extent the scientific criteria reliability and validity are met.

Ferson and Ginzburg (1996) seem to base their work on objectives 1 and 2, as will be clear from the discussion below. The key point being made is that the output of the risk assessment should correspond to the information and knowledge available ("the evidence"). The analysts should not base their assessment on additional assumptions and judgements.

The aim of the assessments in Chapter 5 represents a special case of objective 1, as will be noted in Section 8.3.2.

Inter-subjectivity is the important requirement in 1. We may have situations with experts providing specific probability distributions (reflecting their knowledge), but the assessments would not be inter-subjective. For example, it is common to conduct risk assessments by eliciting input epistemic probabilities from appointed experts (the analysts are not assigning probabilities themselves). Through this process a faithful representation of the information of the experts may be obtained, but it will not in general be inter-subjective. The produced probabilities are subjective knowledge descriptions of the unknown quantities expressing the degrees of beliefs of the experts, in addition to a number of analysts' modelling assumptions. The objective of the assessment is then in line with 2 or 3 (which in this case coincides), but not 1.

The perspective 3 means that the results of the assessments are analysts' (experts') uncertainty descriptions. Through the uncertainty assessments a picture of what is known or not known about a particular issue is created by a group of analysts/experts. Subjective probabilities reflecting degrees of beliefs (judgements) is the most common tool used to describe the uncertainties. The assessments in Chapter 6 are in line with this perspective.

Perspective 2 is also based on subjective assessments, but the basis is "evidence" rather than judgements. The motivation is that the assessment should correspond to the information at hand: specific probability distributions presume the existence of information that is typically not available. The theories (a)–(d) mentioned above are to a large extent motivated by this objective.

In the following we will study these objectives in more detail. Firstly we consider purpose 3: the objective of the risk assessment is to obtain judgements about unknown quantities from a qualified group of people (the analysts/experts). Subjective probabilities are used to describe the uncertainties.

8.3.1 Objective 3 (subjective analyst/expert judgements)

The analysts (and the experts they use in their assesments) are consulted as experts in the field studied and the decision-maker expects them to give their faithful report of the epistemic uncertainties about the unknown parameters θ_1, θ_2 and θ. Firstly, it is known that θ_1 is an integer number between 1 and 5. Hence the assessor can conclude that $\theta_1 \in \{1, 2, \ldots, 5\}$. But how likely is it that $\theta_1 = 1$ compared to $\theta_1 = 5$ (say)? Is it more likely that $\theta_1 = 2$ than $\theta_1 = 4$? And so on. Such questions the analysts are expected to answer, as such judgement would support the decision-making. The decision-maker knows that these judgements are based on some knowledge and some assumptions, and are subjective in the sense that others could conclude differently, but these judgements are still considered valuable as the analysts (and the experts they use in the analysis) have strong competence in the field being studied. The analysts are trained in probability assignments and have no problem in transforming their knowledge into probability figures. Suppose that they conclude by assigning numbers as in Table 8.2.

Table 8.2 *Assigned probabilities for* θ_1

x	$P(\theta_1 = x)$
1	0.1
2	0.2
3	0.4
4	0.2
5	0.1

The probabilities are to be understood as knowledge-based (subjective) probabilities with reference to a standard such as drawing a ball from an urn (refer to Section 2.2). If we assign a probability of 0.4 (say) for an event A (as for $\theta_1 = 3$), the uncertainty (degree of belief) of A to occur is comparable with drawing a red ball from an urn having 10 balls where 4 are red. Hence the assessors state that it is much more likely that $\theta_1 = 3$ compared to $\theta_1 = 1$, a factor of 4. The assignments are judgements based on the assessors' background knowledge, which we denote by K. To show the dependency on K we write $P(A \mid K)$, where A is the event of interest. The background knowledge could be based on hard data and/or expert judgements. Assumptions are also included, for example related to the use of specific models. The background knowledge should be reported along with the assigned probabilities.

Now consider a case with "no or little knowledge" about θ_1. What number then should the assessors assign?

They should also assign numbers in this case, as the decision-maker has consulted them to do so (remember that the objective of the assessment is 3). The decision-maker would like them to make a judgement about the parameter. Given the background knowledge, the assessors may assign the same probability to all values 1, 2, 3, 4 and 5. The judgement is that it is as likely that $\theta_1 = x$ as $\theta_1 = y$. The result is a uniform distribution over the set $\{1, 2, \ldots, 5\}$, i.e. $P(\theta_1 = x) = 0.2$, $x = 1, 2, \ldots, 5$.

The uniform distribution is often referred to as a non-informative distribution, but this is a misleading word as the distribution provides information – the distribution indicates that the assessors consider for example the value $\theta_1 = 1$ to be as likely as $\theta_1 = 2$.

The non-informative distribution is in line with the principle of insufficient reason (Bernoulli, 1713; de Laplace, 1814; Sinn, 1980). The principle says that if there is no reason to believe that out of a set of possible, mutually exclusive events any event is more likely to occur than any other, then one should assign the same probability to all events.

Non-informative distributions are more of a theoretical than practical interest. If the analysts have no information about a quantity, they should re-examine the system: look for relevant data, interview experts, and perhaps perform modelling of the phenomena underlying the value of θ_1.

The parameter θ_1 could represent a physical quantity, for example related to temperature or pressure, but it could also be a chance (frequentist probability); refer to Section 2.2. A chance of an event A, p_A, is an expression of variability in a large population of similar situations. It is defined as the fraction of A events occurring when considering an infinite number of similar situations to the one (those) studied. The variation in the populations that generates the true value of the chance (or the frequentist probability) is referred to as *aleatory (stochastic) uncertainty*; refer to Section 8.2. This uncertainty is, however, not an uncertainty for the analysts (experts), and is better referred to as a variation, as noted in Section 2.2.

This conclusion is in line with Winkler (1996) and Lindley (2000) among others, who view all uncertainties in this context as epistemic uncertainties, a result of lack of knowledge. However, for the purpose of analysing uncertainties and risk it may be useful to introduce models – and variation represents a way of modelling the phenomena studied. In the case of a coin, the model is defined as follows: if we throw the coin over and over again, the fraction of heads will be p. When throwing a die we would establish a model expressing that the distribution of outcomes is given by $(p_1, p_2, \ldots, p_6)$, where p_i is the fraction of outcomes showing i. These fractions are parameters of the models, and they are referred to as frequentist probabilities in a traditional classical statistical setting and as chances in the Bayesian setting, as noted above.

Such models are called probability models or stochastic models. They constitute the basis for statistical analysis, and are considered essential for assessing the uncertainties and drawing useful insights (Winkler, 1996; Helton, 1994). The probability models coherently and mechanically facilitate the updating of probabilities. All analysts and researchers acknowledge the need for decomposing the problem in a reasonable way, but many would avoid the reference to different types of uncertainties as they consider all uncertainties to be epistemic.

Chances (frequentist probabilities) are unknown quantities (fractions in infinite populations) that can be assessed using subjective representations of uncertainty. Probability theory applies to chances (frequentist probabilities) but also to subjective probabilities. Ferson and Ginzburg (1996) acknowledge the importance of probability theory for chances (frequentist probabilities) but seem to discredit subjective probabilities for assessing epistemic uncertainties. Such a stand makes sense, however, if their premise is that probabilities should be objective. Subjective probabilities are not objective, but in line with objective (3) they provide a report on the epistemic uncertainties from a group of people that is supposed to have a strong knowledge about the phenomena and processes being studied.

Propagating the uncertainties through the model

Next we look into the problem of propagating uncertainties about the parameters θ_1 and θ_2 to uncertainties about θ. We may have considerable or scarce information about the parameters. Figure 8.5 presents the subjective probability distribution of the analysts (experts) about θ founded on the subjective probability distributions for θ_1 and θ_2, assuming independence. If we accept the premises for the analysis, the knowledge-based distributions for θ_1 and θ_2, and the independence assumption, most analysts would accept the resulting distribution of θ as a faithful report of the uncertainties about θ_1 and θ_2 since it derived from the rules of probability theory. The marginal distributions of θ_1 and θ_2 we have discussed above; hence it remains to look into the independence assumption. If the analysts (experts) know that $\theta_2 = 5$ (say), would this change the assessors' assignment of $\theta_1 = 1$ (say)? In some cases it would, but in others, it would not. It depends on the situation. The analysts are professional in uncertainty assessments and need to make a judgement about dependence/ independence. If the two parameters θ_1 and θ_2 are associated with completely separate systems or activities, independence may be appropriate.

The same conclusion would also normally be made when considering two similar units when the background knowledge is strong. Say that the distribution in Table 8.2 is based on a considerable amount of relevant data and reflects the variation in some characteristic of this type of unit. Then as an approximation we may use independence as we would not learn much about θ_1 by observing $\theta_2 = 5$ (say).

If, on the other hand, the background knowledge is weak, knowing the value of θ_2 could strongly influence the analyst's judgement about θ_1. A common analysis approach for this type of problem in risk assessment is to assume that $\theta_1 = \theta_2 = \theta_0$, where θ_0 is a common characteristic of the unit, for example a chance representing the fraction of units having a specific property among an infinite population of similar units. Then $\theta = (\theta_0)^2$ and the dependency has been incorporated into the analysis. A subjective probability is used to express the epistemic uncertainties about θ_0 and this leads to a subjective probability about θ. Alternatively, we may assign a number to the conditional probability $P(\theta_1 = x \mid \theta_2 = y)$ or $P(\theta_2 = x \mid \theta_1 = y)$, to produce the simultaneous distribution of θ_1 and θ_2, and from this derive the distribution of θ. The challenge is to find a rationale for determining the conditional probabilities. In addition to modelling, expert judgements are often used to support the probability assignments.

Thus from a theoretical point of view, one may argue that probability theory does provide methods appropriate for propagating epistemic uncertainties. The produced distribution in Figure 8.5 expresses the assessors' uncertainty distribution, given the background knowledge K of the assignments. This background knowledge includes assumptions and suppositions made, models used, etc. Based on this background knowledge, we are led to the distribution of Figure 8.5. This distribution reflects the judgements made by the analysts (experts), conditional on K.

In complex situations, when the propagation is based on many parameters, strong assumptions may be required to be able to carry out the analysis. The analysts may acknowledge a degree of dependencies, but the analysis, which typically is conducted using Monte Carlo simulation, may not be able to describe these in an adequate way. Hence the assessment $P(\theta=x \mid K)$ must be understood and communicated as probabilities conditional on this constraint. The use of subjective probabilities provides a logical consistent basis for assessing and reporting the epistemic uncertainties, but there are "measurement problems" associated with the probability specifications. Nonetheless, the analysis could provide useful decision support: A group of analysts (experts) have reported their qualified judgements about a set of parameters and propagated this through the model (here $\theta_1\theta_2$) to obtain a judgement about the overall system or activity performance. The boundaries and limitations of the assessment are acknowledged, and sensitivity analyses are conducted to reveal the effect of varying critical assumptions.

The uncertainty report is also conditional on the choice of model (here $\theta_1\theta_2$). The model is included in the background knowledge. A model is always wrong as it is a simplified representation of the real world, but it could still be useful for its purpose. The model is introduced to study the performance of the system, to see how parameters affect the overall system or activity performance. Despite the fact that the model is not 100 per cent accurate – there could be a difference between θ and the model output given by $\theta_1\theta_2$ (which is referred to as model inaccuracy or model uncertainty) – the risk assessment could provide useful insights, refer to discussions in Section 6.5.1. Again we have to acknowledge the need for seeing the probabilities in relation to the background knowledge.

The reliability and validity requirements

To relate this discussion to the scientific criteria of reliability and validity we remember from Section 4.3 that reliability is concerned with the consistency of the "measuring instrument" (analysts, experts, methods, procedures) whereas validity is concerned with the success at "measuring" what one sets out to "measure" in the analysis. More precisely we defined

Reliability: The extent to which the risk analysis yields the same results when repeating the analysis (R).

Validity: The degree to which the risk analysis describes the specific concepts that one is attempting to describe (V).

Depending on the objectives of the analyses, more specific and detailed interpretations (sub-criteria) of the above general definitions of reliability and validity can be formulated:

Reliability

- The degree to which the risk analysis methods produce the same results at reruns of these methods (R1).
- The degree to which the risk analysis produces identical results when conducted by different analysis teams, but using the same methods and data (R2).
- The degree to which the risk analysis produces identical results when conducted by different analysis teams with the same analysis scope and objectives, but no restrictions on methods and data (R3).

Validity

- The degree to which the assigned probabilities adequately describe the assessor's uncertainties of the unknown quantities considered (V2).
- The degree to which the epistemic uncertainty assessments are complete (V3).
- The degree to which the analysis addresses the right quantities (V4).

The criterion V1: the degree to which the produced risk numbers are accurate compared to the underlying true risk is not relevant in the context discussed here (there is no true risk).

As argued in Chapter 6, and summarised in Section 6.5.3, the reliability and validity criteria are to a large extent met when the assessments are adequately conducted. However, several difficulties are identified. For the reliability criteria a main problem is the fact that the background knowledge that the assignments are based on would not be exactly the same from analysis to analysis. However, if the methods and data are fixed, the differences from one analysis to another are not likely to be large if V2 is met.

The validity requirements could also be questioned:

- important uncertainty factors may be hidden in the background knowledge (V2,V3)
- the uncertainty assessments may not be complete (V3).

Hence, for the analysis to meet the validity criteria, it is essential that the background knowledge is reported along with the assigned probabilities. We refer to Chapter 6.

8.3.2 Objectives 1 (inter-subjective descriptions) and 2 (faithful uncertainty representation)

Assume now objective 1; the aim of the risk assessment is to obtain an "inter-subjective" description of the unknown quantities. Suppose that chance distributions (frequentist distributions) exist and are known (can be accurately estimated). These distributions we denote by p_1, p_2 and p. For example, $p_1(2)$ equals $P_f(\theta_1 = 2)$, where P_f is a chance (frequentist) probability. The distribution p(x) reflects the aleatory uncertainties. Then provided that θ_1 and θ_2 are independent, we can calculate the true distribution of θ, i.e. p. The assessment is in line with objective 1.

Next we consider situations where the information basis is smaller and we cannot establish accurate chance distributions. What should we do then?

Assume that we know that $\theta_1 \in \{1, 2, \ldots, 5\}$, but we have little information about the chance distribution $p_1(x)$, $x = 1, 2, \ldots, 5$. A chance distribution could also be hard to define. We can make the same type of characterisation for θ_2, and consequently the basis for assessing θ is poor. How should we then faithfully report the uncertainties?

The answer by Ferson and Ginzburg (1996) is to use interval analysis: as $\theta_i \in \{1, 2, \ldots, 5\}$ we can conclude that the product $\theta = \theta_1\theta_2$ lies in the interval $[1, 25]$, i.e. $\theta \in \{1, 2, \ldots, 25\}$. More precisely, we can conclude that $\theta \in \{1, 2, 3, 4, 5, 6, 8, 9, 10, 12, 15, 16, 20, 25\}$ as it is clear that θ cannot take any of the values 7, 11, 13, 14, 17, 18, 19, 21, 22, 23, 24. This analysis is trivial – it produces objective results in the sense that we all agree that θ is in this interval under the given assumptions.

This approach meets objective 1, to obtain an "objective" ("inter-subjective") description of the unknown quantities. It follows from the analysis that $\theta \in \{1, 2, 3, 4, 5, 6, 8, 9, 10, 12, 15, 16, 20, 25\}$, but any probability number for these values of θ has no rigorous basis and should therefore not be reported. The assessment should not be based on subjective assessments not supported by the available information.

But such a perspective can be challenged. The decision-maker may not be satisfied with such an analysis only. He/she would expect the analysts (experts) to be able to make some statements about likelihoods: is it as probable that θ is as high as it is low? Are there values of θ that are considered quite unlikely compared to other values? And so on. However, this would mean a shift from objective 1 to 3.

We may also see a shift to objective 2. The analysts (experts) may not be willing or able to assign subjective probabilities that are more precise than this:

$$P(1 \leq \theta_1 \leq 5) = 1,$$

$$1/2 \leq P(\theta_1 = 3) \leq 1.$$

Similar assignments are given for θ_2. From this we can obtain bounds for $P(\theta_1 = 3, \theta_2 = 3)$. However, non-trivial bounds are not that easily derived without making additional assumptions concerning the knowledge about the dependencies between θ_1 and θ_2. For example, if bounds have been established for $P(\theta_2 = 3 \mid \theta_1 = 3)$, say

$$1/2 \leq P(\theta_2 = 3 \mid \theta_1 = 3) \leq 1,$$

it follows that

$$1/4 \leq P(\theta_1 = 3, \theta_2 = 3) \leq 1,$$

This would be an assessment in line with objective 2): a faithful representation of the uncertainties reflecting the information and knowledge available.

Let us consider a somewhat more realistic example. A unit is considered for possible use in a process plant. The unit is based on new technology and a risk assessment is conducted at an early stage of the planning process where test results are lacking. A key uncertainty factor is how the technology would work. We then have lack of support for a specific distribution for the performance of the system. In the risk assessment we could, however, make an assumption, for example that the unit would not work as planned in a maximum of 10 per cent of the situations – the unreliability does not exceed 10 per cent. The rationale for this assumption is that the technology would not be used if a performance level below 10 per cent is achieved in testing of the unit. A reference is made to other similar units (not based on the same technology but performing a similar function).

A system is defined comprising two such units. The system is viewed as a parallel system of the units, i.e. the system is functioning if at least one of the units is functioning. Let θ_0 be the unreliability of a unit, i.e. the chance (frequentist probability) that the unit is not functioning. As a model of the unreliability θ of the system, we use the model $\theta = (\theta_0)^2$, which is based on the assumption of independence. Based on this model and the assumption that $\theta_0 \leq 0.10$, we obtain an upper level of the system unreliability: $\theta \leq 0.01$.

However, the model $\theta = (\theta_0)^2$ can be disputed. If no assumption is made on the dependencies of the units, this leads to a high upper bound on the system unreliability as will be explained in the following. Let X_i be the state of unit i, defined as 1 if it is not functioning and 0 otherwise, and let Y denote the state of the system defined analogously. We assume $Y = X_1 \cdot X_2$. We have

$P_f(X_i = 0) = \theta_0$ and $P_f(Y=0) = \theta$, where the P_fs are chances (frequentist probabilities). Hence, in the case of independence between X_1 and X_2, we obtain

$$\theta = P_f(X_1 = 0)\, P_f(X_2 = 0) = (\theta_0)^2.$$

In the general case,

$$\theta = P_f(X_1 = 0)\, P_f(X_2 = 0 \mid X_1 = 0) \le 0.1 \cdot P_f(X_2 = 0 \mid X_1 = 0).$$

It is difficult, in practice impossible without making assumptions, to find an upper bound on $P_f(X_2 = 0 \mid X_1 = 0)$ that is lower than 1. The frequencies of dependent events are difficult to accurately predict. The conclusion is that the bound 1 has to be used; the result being that we arrive at the system unreliability bound $\theta \le 0.1$.

In many cases this upper limit is, however, considered rather non-informative. A system unreliability level of 10 per cent is often not accepted, and the decision-maker would ask for a more refined assessment. It is expected that the analysts are able to give some qualified judgements about the unit dependencies, so that an overall more precise assessment of the system performance can be reported. The result could be more detailed modelling of the system and the dependencies, and/or input from experts. Again we are led to an objective of the assessment closer to 3.

Ferson and Ginzburg (1996) suggest a combined probability analysis and interval analysis (a probability-bound analysis). For the components where the aleatory uncertainties cannot be accurately estimated, interval analysis is used. In this way uncertainty propagation is carried out in the traditional probabilistic way for some components, and intervals are used for others. A framework is established for treatment of both types of uncertainties. To understand the set-up analysed it is essential that we clarify the types of probabilities and analyses conducted:

(a) For parameters θ_i where the aleatory uncertainties cannot be accurately estimated, use interval analysis expressing that $a_i \le \theta_i \le b_i$ for constants a_i and b_i.
(b) For parameters θ_i where the aleatory uncertainties can be accurately assessed, use probabilities (frequentist probabilities) to describe the distribution over θ_i.
(c) Combine (a) and (b) to generate a probability distribution over θ, for the different interval limits. For example, assume that for $i=1$; interval analysis is used with bounds a_1 and b_1, whereas for $i=2$, a probabilistic analysis is used. Then we obtain a probability distribution over θ when $\theta_1 = a_1$ and a probability distribution over θ when $\theta_1 = b_1$.

Depending on the basis for the intervals, the analysis is in line with objective 1 or 2. The approach does not provide the decision-maker with specific analyst and expert judgements about epistemic uncertainties. In many cases we do not have situations as in (b), where the aleatory uncertainties can be accurately assessed, and then the analysis reduces to an interval analysis. On the other hand, if accurate estimates of the aleatory distributions can be obtained, the analysis is purely probability-based.

The critical issue is the use of knowledge-based (subjective) probabilities to assess the epistemic uncertainties. Some researchers and analysts argue that the subjective probabilities cannot be used to treat situations where there is less information than a single probability distribution requires (Dubois, 2010). But such a statement is not acknowledging what a subjective probability means: expressing the assessors' degree of belief. A subjective probability can always be assigned, regardless of the information available. The problem is rather that these researchers and analysts would like to see a stronger rationale for the probabilities. We are again led to a discussion about the objective of the assessment.

Example: Two prospects with different chance distributions

Consider the example introduced by Huber (2010) (see also Aven, 2010d): a decision-maker has asked a risk analyst for advice. She has two prospects to choose from, both risky. Each offers an opportunity to gain €6.

For prospect A there is a 1/6 chance of success.

For prospect B the decision-maker is informed that the chance of success p (winning €6) is somewhere between 0 and 2/3. These bounds are based on theoretical limits, not on any data or experience. According to a standard Bayesian analysis, using a uniform distribution over [0, 2/3], this implies that the (predictive) subjective probability of wining €6 is

$$P(€6) = \int_{[0,2/3]} P(€6|p) \ 3/2 \ dp = \int_{[0,2/3]} p3/2 \ dp = 1/3.$$

Hence, the probability of winning is a factor two higher than for prospect A, where the chance of winning €6 is 1/6. Just by comparing the probabilities of winning, the decision-maker is led to prospect B.

If we have not one opportunity for prospect B, but three, then the probability of winning at least once becomes $1- (1-p)^3$ if p were known. However, p is unknown and again, using the uniform distribution for the chance p, we obtain the (predictive) subjective probability:

$$P(\text{winning at least once}) = 1 - P(\text{not winning})$$

$$= 1 - \int_{[0,2/3]} (1 - p)^3 \ 3/2 \ dp = 1 - 10/27 = 17/27 \approx 0.63.$$

For three opportunities for prospect A, the corresponding probability becomes $1 - (5/6)^3 = 91/216 \approx 0.42$. It seems that the analysts judge prospect B to be better than prospect A, as B has the highest probability of winning.

According to Huber (2010), it appears that the decision-maker's interests are not well served by the probabilistic analysis offered by the risk analyst. The decision-maker would acknowledge a probability of winning equal to 0.63, but the actual chance could be close to nil.

However, the above rationale makes sense if you agree with the prior distribution. As noted above, the uniform distribution is not, in fact, a non-informative distribution. The analyst expresses, for example, that it is as likely that the chance is below $1/9$ as it is in the interval $[5/9, 6/9]$. If the assessment is according to objective 3, the analysts should make their judgement about the chance and then compute the predictive probability as indicated by the calculations above. However, the decision-maker may not be satisfied with this subjective risk description. The analysts' judgements may lead to poor predictions if they have poor background knowledge about the prospects. And this is a key point in the discussion. Most risk experts would agree, that a complete risk description should inform about the chance interval for prospect B: $[0, 2/3]$, which means that the chance interval for say three opportunities is $[0, 0.70]$ (note that $1 - (2/3)^3 \approx 0.70$). However, as stressed above, we also need risk assessments according to objective 3 to support the decision-making. The decision-maker consults experts in the field to be informed by their likelihoods (degrees of belief) and the resulting risk picture based on their knowledge. This picture has to be appreciated for what it is, a knowledge-based judgement about unknown quantities, not an objective risk description. Sensitivity analysis is needed to show how the output probabilities depend on key assumptions and the assigned input probabilities. Special focus should be on the possible small chance values.

If chances and chance distributions can be established (justified), a full risk description needs to assess uncertainties about these quantities. To meet the validity criterion, it would not be sufficient to provide predictive distributions alone, as important aspects of the risk then would not be revealed. The predictive distributions would not distinguish between the variation and the epistemic uncertainties. Focus is not on the right quantities. Dubois (2010) expresses the problem in this way: if the ill-known inputs or parameters to a mathematical model are all represented by single probability distributions, either objective when available or subjective when less information is available, then the resulting distribution on the output can hardly be properly interpreted: "the part of the resulting variance due to epistemic uncertainty (that could be reduced) is unclear". However, as this discussion makes clear, this indicated inadequacy of the subjective probabilities for reflecting uncertainties is more an issue of addressing the right quantities: if chances can be established (justified) the subjective probabilities should be used to reflect the uncertainties about these chances.

Huber (2010) seems to indicate that it is possible to assign wrong prior distributions (Section 2.2). He writes: "When a prior distribution is wrong and few data are available, decisions are likely to be bad." This way of speaking is not in line with the subjective Bayesian perspective. A subjective probability expresses the assigner's uncertainty (degree of belief) given his/her knowledge – there exists no true probability distribution. Of course, the choice of a prior distribution can, to varying degrees, support accurate predictions of observables, but that is not the same as saying that the prior is wrong or right. The problem raised by Huber is a more general one, the estimation (prediction) accuracy of risk assessments. But is such accuracy a requirement to ensure quality of a risk assessment? Think of the analysis of rare events. How can one "guarantee" accurate estimates? The question again relates to the aim of the risk assessment. In line with objectives 2) and 3), risk assessment is more about uncertainty descriptions than accurate estimation.

As mentioned above, several alternative theories to probability have been established, including the theories of possibility and evidence. Below we outline the main ideas of the former theory (Dubois, 2006; Aven and Zio, 2011). These theories can be seen in line with objective 2). Their motivation is that the intervals produced correspond better to the information available.

Possibility theory

In possibility theory uncertainty is represented by using a possibility function $r(x)$. For each x in a set Ω, $r(x)$ expresses the degree of possibility of x. When $r(x) = 0$ for some x, it means that the outcome x is considered an impossible situation. When $r(x) = 1$ for some x, it means that the outcome x is possible, i.e. is just unsurprising, normal, usual (Dubois, 2006). This is a much weaker statement than when probability is 1.

The possibility function r gives rise to probability bounds, upper and lower probabilities, referred to as the necessity and possibility measures (Nec, Pos). They are defined as follows:

The possibility (plausibility) of an event A, Pos(A), is defined by

$$Pos(A) = \sup_{\{x \in A\}} r(x), \tag{8.1}$$

and the necessity measure Nec(A) is defined by $Nec(A) = 1 - Pos(\text{not } A)$.

Let $\mathscr{P}(r)$ be a family of probability distributions such that for all events A,

$$\text{Nec}(A) \leq P(A) \leq \text{Pos}(A).$$

Then

$$\text{Nec}(A) = \inf P(A) \quad \text{and} \quad \text{Pos}(A) = \sup \quad P(A) \qquad (8.2)$$

where inf and sup are with respect to all probability measures in $\mathscr{P}$. Hence the necessity measure is interpreted as a lower level for the probability and the possibility measure is interpreted as an upper limit. Using subjective probabilities, the bounds reflect that the analyst (expert) is not able or willing to precisely assign his/her probability. He or she can only describe a subset of $\mathscr{P}$ which contains his/her probability (Dubois and Prade, 1989).

A typical example of possibilistic representation is the following (Anoop and Rao, 2008; Baraldi and Zio, 2008): we consider an uncertain parameter x. Based on its definition we know that the parameter can take values in the range [1, 3] and the most likely value is 2. To represent this information, a triangular possibility distribution on the interval [1, 3] is used, with maximum value at 2; see Figure 8.6.

From the possibility function we define α cut sets $F_\alpha = \{x: r(x) \geq \alpha\}$, for $0 \leq \alpha \leq 1$. For example $F_{0.5} = [1.5, 2.5]$ is the set of x values for which the possibility function is greater than or equal to 0.5. From the triangular possibility distribution in Figure 8.6, we can conclude that if A expresses that the parameter lies in the interval [1.5, 2.5], then $0.5 \leq P(A) \leq 1$.

From (8.1) and (8.2) we can deduce the associate cumulative necessity/ possibility measures $\text{Nec}(-\infty, x]$ and $\text{Pos}(-\infty, x]$ as shown in Figure 8.7. These measures are interpreted as the lower and upper limiting cumulative probability distributions for the uncertain parameter x. Hence the bounds for the interval [1, 2] are $0 \leq P(A) \leq 1$.

These bounds can be interpreted as for the interval probabilities: the interval bounds are those obtained by the analyst (expert) as he/she is not able or willing to precisely assign his/her probability – the interval is the best he/she can do given the information available.

The scientific criteria reliability and validity

The motivation for objective 2 is to obtain a faithful representation of the information and knowledge available. In this sense an analysis in line with this objective would to a large extent meet the reliability and validity criteria. For example, the criterion R2: "the degree to which the risk analysis produces identical results when conducted by different analysis teams, but using the same methods and data" would be easier to meet than using subjective probabilities to express uncertainties. However, one may question to what

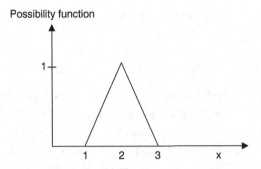

Figure 8.6   Possibility function for a parameter on the interval [1, 3], with maximum value at 2.

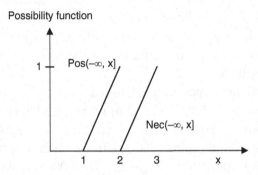

Figure 8.7   Bounds for the probability measures based on the possibility function in Figure 8.6.

extent the validity criterion is satisfied: "the degree to which the risk analysis describes the specific concepts that one is attempting to describe" (V).

   Producing an interval ([0.2, 0.6], say) for the subjective probability P(A), the analysts (experts) are not able or willing to precisely assign their probability P(A). The decision-maker may, however, request that the analysts (experts) make such assignments – the decision-maker would like to be informed by the analysts' (experts') degree of belief as discussed in Section 8.3.1 (Aven, 2010b).

8.3.3 Discussion

Seeing risk assessment as an aid for decision-making, alternative approaches for the representation and treatment of uncertainties in risk assessment are required. Different approaches provide a broader and more informative decision basis than one approach alone. A Bayesian analysis without thorough considerations of the background knowledge and associated assumptions would normally fail to reveal important uncertainty factors. Such considerations (qualitative assessments) are essential for ensuring that the decision-makers are not seriously misled by the risk assessment results.

It is a huge step from such assessments to methods that quantitatively express, and bound, the imprecision in the probability assignments. These methods are also based on a set of premises and assumptions, but not to the same degree as the pure probability-based analyses. Their motivation is that the intervals produced correspond better to the information available. However, as discussed in the previous section, the decision-maker may request that the analysts (experts) make more specific assignments reflecting their degrees of belief.

Following this view, we should continue to conduct probability–based analysis reflecting the analysts' degrees of belief about unknown quantities, but we should also encourage additional assessments. These include sensitivity analyses to see how sensitive the risk indices are with respect to changes in basic input quantities, for example assumptions and suppositions (Saltelli *et al.*, 2008; Helton *et al.*, 2006), but also crude qualitative assessments of uncertainty factors as discussed in previous chapters. The use of imprecision intervals would further point at the importance of key assumptions made.

It is raised as a research goal to establish a unifying framework for representation and treatment of uncertainties in risk assessment (Aven and Zio, 2011). Such a framework can be based on the ideas outlined in the previous sections. Imprecision intervals would constitute an integral part of such a framework. To make these intervals meaningful in a practical decision-making context, proper interpretations are required.

The underlying theories are not reported in detail here, as they are technical and not important for the proper understanding of the results. We believe that to make the alternative approaches operational in a practical decision-making context, we should leave out the technical terminology used in these theories. If one looks at various attempts that have been made to use alternative representations of uncertainty in risk assessment contexts, the general impression is that they are extremely difficult to understand and appreciate. We believe that in a practical decision-making context they would typically be rejected as they add more confusion than insights.

Many researchers and analysts are sceptical about the use of "non-probabilistic" approaches (such as those of the four categories (a) – (d) listed at the beginning of Section 8.3) for the representation and treatment of uncertainty in risk assessment for decision-making. An imprecise probability result is considered to provide a more complicated representation of uncertainty (Lindley, 2000). By arguing that the simple should be favoured over the complicated, Lindley (2000) takes the position that the complication of imprecise probabilities seems unnecessary. In a more rejecting statement, Lindley (2006) argues that the use of interval probabilities goes against the idea of simplicity, as well as confuses the concept of interpretation with the practice of measurement procedures. The standard, Lindley (2006) emphasises (for example the use of an urn for randomly drawing balls), is a conceptual comparison. It provides a norm, and measurement problems may make the assessor unable to behave according to it. Bernardo and Smith (1994, p. 32) call the idea of a formal incorporation of imprecision into the axiom system "an unnecessary confusion of the *prescriptive* and the *descriptive*" for many applications, and point out that measurement imprecision occurs in any scientific discourse in which measurements are taken. They make a parallel with the inherent limits of a physical measuring instrument, where it may only be possible to conclude that a reading is in the range 3.126 to 3.135, say. Then, we would typically report the value 3.13 and proceed as if this were the precise number:

We formulate the theory on the prescriptive assumption that we aspire to exact measurement . . ., whilst acknowledging that, in practice, we have to make do with the best level of precision currently available (or devote some resources to improving our measuring instruments!)

(Bernardo and Smith, 1994, p. 32)

Many analysts argue fiercely for a strict Bayesian analysis. A typical statement is (North, 2010):

For me, the introduction of alternatives such as interval analysis to standard probability theory seems a step in the wrong direction, and I am not yet persuaded it is a useful area even for theoretical research. I believe risk analysts will be better off using standard probability theory than trying out alternatives that are harder to understand, and which will not be logically consistent if they are not equivalent to standard probability theory.

However, as argued in this paper, this approach does not solve the problems raised. The decision basis cannot be restricted to subjective probabilities: there is a need to go beyond the Bayesian approach.

In the end, any method of uncertainty representation and analysis in risk assessment must address a number of very practical questions before being applicable in support of decision-making:

• How completely and faithfully does it represent the knowledge and information available?
• How costly is the analysis?

- How much confidence does the decision-maker gain from the analysis and the presentation of the results? (This opens up the issue of how one can measure such confidence.)
- What value does it bring to the dynamics of the deliberation process (the managerial review and judgement process)?

More so, any method which intends to complement, or in some justified cases supplement, the commonly adopted probabilistic approach to risk assessment should demonstrate that the efforts needed for the implementation and familiarisation, by the analysts and decision-makers, are feasible and acceptable in view of the benefits gained, in terms of the above questions and, eventually, of the confidence in the decision made (Aven and Zio, 2011).

In this book we have highlighted probability as a measure of uncertainty and the recommended risk description (A,C,U,P,K) reflects this. However, the (A,C,U) risk perspective is general and applicable also for alternative approaches and theories for representing and expressing uncertainties in a risk assessment context. Risk exists as (A,C,U) independent of the perspectives adopted. To describe the risk, different tools could be used: probability, possibility measures, etc. The risk description then takes the form (A,C,Q,K), where Q expresses the measure used to represent the uncertainties U, and K is the information and knowledge that Q is based on. In this perspective probability models may be introduced whenever considered appropriate, to structure and ease the uncertainty assignments. It is, however, beyond the scope of this book to further discuss these alternative risk descriptions.

9

Conclusions

In this book we have investigated to what extent risk assessments and in particular quantitative risk assessments, meet the scientific quality requirements of reliability and validity. While reliability is concerned with the consistency of the "measuring instrument" (analysts, experts, methods, procedures), validity is concerned with the success at "measuring" what one set out to "measure" in the analysis. For each of these main criteria a set of sub-criteria are defined.

To be able to perform these investigations we have to study the scientific building blocks of the assessments (related to probability, risk, uncertainty, models, etc.), but also the role of the assessments in the decision-making process. What type of decision support should the assessments provide? Are the objectives (expectations) accurate risk estimates and/or uncertainty characterisations? The scientific quality of the assessments obviously needs to be seen in relation to the objectives. Also the requirements of reliability and validity depend on these objectives. Using these criteria, we evaluate the quality of the assessments for different objectives of the assessments. Three case studies are used to illustrate the analysis. The first of these examples is related to the analysis of accident data, the second relates to the siting of an LNG plant and the third discusses the design of a safety system.

The investigation shows that the reliability and validity criteria are not in general satisfied. Under certain conditions the criteria are met, and to a varying degree depending on the risk perspective and aim of the risk assessment.

The traditional statistical methods meet the reliability and validity criteria only if a large amount of relevant data is available. If the objective of the risk assessment is to express uncertainties, the reliability and validity criteria are to a large extent met provided that the assessments are properly conducted. However, several problems have been identified, including:

The background knowledge that the assignments are based on need not be
    exactly the same from analysis to analysis (R1–R3).

Important uncertainty factors may be hidden in the background know-
    ledge (V2,V3).

The uncertainty assessments may not be complete (V3).

In addition, the V4 criterion is questioned when the risk assessment has a
focus on the estimation of fictional parameters of probability models. The
idea is that the parameter reflects some underlying property of the activity
studied, but often this parameter cannot be given a meaningful interpretation.
Nonetheless the assessment is conducted, as experience shows that it can be a
useful tool for practical analysis. However, this type of pragmatism cannot be
accepted if we require that our assessment should be built on a scientific basis.
Then all quantities introduced should be properly defined and allow for
meaningful interpretations.

The V4 criterion is also problematic if "non-probabilistic" approaches are
used to express uncertainties. These approaches are based on bounds for the
epistemic-based probabilities. Specific probability assignments are not in
general used as they cannot be given a rigorous basis. These approaches
may improve on the reliability criteria compared to the Bayesian approaches,
but commonly fail when it comes to the validity criterion as they do not
adequately inform the decision-maker, as discussed in Section 8.3. Expressing
epistemic uncertainties means subjective assignments but decision-making
normally needs to be supported by qualified judgements. Sensitivity analysis
is considered an important tool to show how the results are dependent on the
assumptions made.

We argue that risk assessment is a science in its own right. Risk assessment
cannot and should not be judged by reference to the traditional science
paradigms alone, such as the natural sciences, social sciences, mathematics
and probability theory. We may view the science of risk assessment as

the development of concepts, principles, methods and models to identify, analyse and
evaluate (assess) risk, in a decision-making context (Aven, 2004).

This science, which is founded on the international scientific journals in the
field, provides the basis for the scientific method of risk assessment. In the
book we show that such a method can be established, in some cases and under
some conditions. The best framework for this method is in our view the (A,C,
U) risk perspective, as it the most general perspective and not dependent on
the construction of fictional model parameters. The scientific method of risk
assessment based on this perspective can be summarised as follows:

1. We define risk by the two-dimensional combination of
   (i) events, A, and the consequences of these events, C, and
   (ii) the associated uncertainties, U (will A occur and what value will C take?).

   Often the A events are specified, for example as gas leakages in a process plant or as terrorist attacks in a country, but we may also allow for new types of such events, a new type of virus for instance. We speak then often about "unknown uncertainties" ("unknown unknowns", ignorance or non-knowledge) – we do not know what we do not know, in contrast to "known uncertainties" ("known unknowns") – we know what we do not know.

2. To describe the uncertainties, we use subjective (knowledge-based) probabilities. If the probability equals 0.1 (say), this means that the assessor compares his/her uncertainty (degree of belief) about the occurrence of the event with the standard of drawing at random a specific ball from an urn that contains 10 balls.

3. A risk description based on this perspective includes the following elements: (A,C,U,P,K), that is, risk is described by events and consequences, knowledge-based probabilities P, assessment of uncertainties U, and K the background knowledge that U and P are based on. The assessed uncertainties and probability assignments are based on hard data, expert judgements and models. The U component may for example be a qualitative assessment of uncertainty factors (assumptions that the probabilities are based on). It partly relates to the unknown unknowns.

4. This perspective acknowledges that risk extends beyond probabilities. Probability is just a tool used to express the uncertainties, but it is not a "perfect" tool.

5. Stochastic models (with parameters) expressing aleatory uncertainty, i.e. variation in populations of similar units, are used to ease the probability assignment. The probability models make it possible to coherently and mechanically facilitate the updating of probabilities in line with the Bayesian paradigm. However, such models need to be justified, and if introduced they are to be considered as tools for assessing the uncertainties about A and C. The estimation of the parameters of the models needs not be the end product of the analysis as in a traditional parametric risk analysis.

6. If a probability model is justified, the uncertainties are not scientific although the epistemic uncertainties about the parameters of the model could be large. The precautionary principle applies when the uncertainties are scientific which means that it is difficult to establish an accurate prediction model for the consequences.

7. A distinction is made between model inaccuracies (the differences between Z and G(X), see Section 6.5.1), and model uncertainties due to alternative plausible hypotheses on the phenomena involved. Model inaccuracy is not reflected in the resulting knowledge-based (subjective) probabilities: the models are included in the background knowledge of the assigned probabilities. The assumptions supporting a model can give rise to uncertainty factors.

8. A risk assessment is a methodology designed to determine the nature and extent of risk, i.e. assess the risk (A,C,U). It comprises the following main steps:
   (a) identification of hazards/threats/opportunities (sources)
   (b) cause and consequence analysis, including analysis of vulnerabilities
   (c) risk description, using probabilities and expected values
   (d) identification and assessments of uncertainty factors
   (e) risk evaluations, i.e. comparisons with possible risk tolerability (acceptance criteria).

9. The results of risk assessments are used to support decision-making. The risk assessment results inform the decision-makers. The decision emerges from a managerial review and judgement phase during which the management/decision-makers consider
   (i) the formal results of risk assessments and other type of assessments
   (ii) the premises, assumptions and limitations of these assessments
   (iii) other issues not captured by the assessments.

10. The decision-making encompasses considerations and weighting with respect to uncertainty (the weight given to the cautionary and precautionary principles) and values. It is the responsibility of the decision-maker to undertake such considerations and weighting, and to make a decision that balances the various concerns.

11. The decision-makers reduce risk by implementing cautious policies (including traditional engineering approaches of "defence-in-depth" to bound the uncertainties).

12. If risk criteria in the form of risk acceptance limits (such as IR < 0.001) are introduced, they have to be used as nothing more than reference levels to inform the decision-maker, not to provide a mechanical procedure for what is an acceptable or unacceptable risk.

Appendix A

Introduction to probability theory and statistical analysis

This appendix reviews some elementary probability theory and statistical analysis. The presentation is based on Aven (2003, 2008a).

A.1 Probability theory

A.1.1 The meaning of a probability

The probability of an event A, P(A), can be defined in different ways. It is common to distinguish between three types of probabilities, or more precisely, three types of interpretations:

- classical probabilities
- frequentist probabilities (relative frequency-interpreted probabilities)
- subjective (knowledge-based) probabilities.

The classical interpretation applies only in situations with a finite number of outcomes which are equally likely to occur. According to the classical interpretation the probability of A is equal to the ratio between the number of outcomes resulting in A and the total number of outcomes, i.e.

P(A) = Number of outcomes resulting in A/Total number of outcomes.

As an example consider the tossing of a die. Here P(the die shows one) = 1/6 since there are six possible outcomes which are equally likely to appear and only one that gives the outcome 1.

Following the relative frequency interpretation, probability is defined as the fraction of times the event A occurs if the situation considered were repeated (for real or hypothetically) an infinite number of times. Thus, if an experiment is performed n times and the event A occurs n_A times, the P(A) is equal to the limit of n_A/n as n goes to infinity, i.e. the probability of the event A is the limit of the fraction of the number of times event A occurs when the number of experiments increases to infinity. Note that a classical inter-preted probability is equal to a relative frequency-interpreted probability. For example in the die example above, the proportion of times the die show one is 1/6 in the long run and hence the relative frequency-interpreted probability is 1/6.

In most real-life situations, the relative frequency-interpreted probability is unknown and has to be estimated from experience data. To illustrate, let us look at an example.

We consider a fire detector of a type D. The function of the detector is to raise the alarm at a fire. Let A denote the event "the detector does not raise the alarm at a fire". To find $P(A)$, assume that tests of n detectors of type D have been carried out and the number of detectors that are not functioning, n_A, is registered. As n increases, the fraction n_A/n will be approximately constant and approach a certain value (this fact is called the strong law of large numbers). This limiting value is called the probability of A, $P(A)$. Thus if for example $n = 10000$ and we have observed $n_A = 50$, then $P(A) \approx 50/10000 = 5/1000 = 0.005$ (0.5%). Note that a probability is by defini-tion a number between 0 and 1, but is also often expressed in percentages.

Following the subjective interpretation, $P(A)$ is a subjective measure of uncertainty. This means that we (who assign the probability) compare the uncertainty of event A occurring with drawing a favourable ball from an urn with $P(A) \times 100\%$ favourable balls under standard experimental conditions. This means that we have the same degree of belief in the event A occurring as drawing a favourable ball from an urn with $P(A) \times 100\%$ favourable balls.

All subjective probabilities are conditioned on some background know-ledge K, say. Thus a more precise way of writing the probability $P(A)$ is $P(A|K)$, which is the common way of expressing a conditional probability. To simplify the writing we normally omit the K. This should not cause any problem as long as the background knowledge is fixed throughout the argumentation.

To understand the concept of a subjective probability let us look at an example. Let A represent the event that a patient develops an illness S over the next year, when the patient shows symptoms V. We do not know if A will occur or not – there is uncertainty associated with the outcome. However, we can have an opinion on how likely it is that the patient will develop the illness. Statistics show that about 5 out of 100 patients develop this illness over the course of one year, if they show the symptoms V. Is it then reasonable to say that the probability that A will occur is equal to 5 per cent? Yes, if this is all the knowledge that we have available, then it is reasonable to say that the probability that the patient will become ill next year is 0.05 if the symptoms V are present. If we have other information relevant about the patient, our probability can be entirely different. Imagine, for example, that the particular patient also has an illness B, and that his/her general condition is somewhat weakened. Then it would be far more likely that the patient develops illness S. The physician who assesses the patient may, perhaps, set a probability of 75 per cent for this case: for four such cases that are relatively similar, he/she predicts that three out of these will develop the illness.

Based on this line of thought, a correct or true probability does not exist. Even if one throws a die, there is no correct probability. This may seem strange, but one must differentiate between proportions, observed or imaginary, as in relative frequencies and a subjective probability as considered here. Imagine throwing a die a great many times – say, 6000 times. We would then obtain (if the die is "normal") about 1000 showing a "1", about 1000 showing "2", and so on. In the population of 6000 throws, the distribution will be rather similar to 1/6 for the various numbers. But, imagine that we did an infinite number of attempts. Then the theory says that we would obtain exactly 1/6. However, these are proportions, observed and resulting from imaginary experiments. They are not subjective probabilities. A subjective probability applies to a defined event which we do not know will occur or not, and which is normally associated with the future. We will throw a die. The die can show a "4", or it can show a different number. Prior to casting the die, one can express one's belief that the die will show a "4". As a rule, this probability is set at 1/6, because it will yield the best prediction of the number of "fours" if we make many throws.

However, there is nothing automatic in our assignment of the probability 1/6. We have to make a choice. We are the ones who must express how likely it is to obtain a "6", given our background knowledge. If we know that the die is fair, then 1/6 is the natural choice. However, it is possible that one is convinced that the die is not fair, and that it will give many more "fours" than usual. Then P("fours") = 0.2 may perhaps be set. No one can say that this is wrong, even though, afterwards one checks the proportion of "fours" for this die and finds it to be fair. When one originally assigned the probability, the background knowledge was different. Probability must always be seen in relation to the background knowledge.

A.1.2 Probability calculus

The rules for probabilities are widely known. We will not repeat them all here, but will only summarise briefly some of the most important ones, understanding P as a subjective probability. The reader is referred to text-books on probability theory.

Probabilities are numbers between 0 and 1. If the event A cannot occur, then $P(A) = 0$, and if A is considered certain, then $P(A) = 1$. If we have two events A and B, then the following formulae hold:

$$P(A \text{ or } B) = P(A) + P(B) - P(A \text{ and } B)$$

$$P(A \text{ and } B) = P(A)\, P(B|A). \tag{A.1}$$

Here $P(B|A)$ represents our probability for B when it is known that A has occurred. If A and B are independent, then $P(B \mid A) = P(B)$; in other words, the fact that we know that A has occurred does not affect our probability that B will occur.

Imagine that we want to express the probability that two systems will both fail. In other words, we will determine $P(A_1 \text{ and } A_2 \mid K)$ where A_1 represents the event that system 1 fails, and A_2 represents the event that system 2 fails. We base our analysis on the assignments

$$P(A_1|K) = P(A_2|K) = 0.05.$$

Is then

$$P(A_1 \text{ and } A_2|K) = P(A_1|K)\, P(A_2|K) = 0.05^2 = 0.25\%?$$

The answer is "yes" if A_1 and A_2 are independent. But are they? If it was known to us that system 1 has failed, would it not alter our probability that system 2 would fail? Not necessarily, it depends on what our background knowledge is,

- what is known to us initially,
- whether there is a "coupling" between these systems in some way or another, for example the failure of one system increases the stress on the other.

If the systems are of the same type and our background knowledge is very small, knowledge that system 1 has failed provides information to us about system 2. In practice, however, we may have so much knowledge about this type of system that we can ignore the information that is associated with A_1. We then obtain independence since

$$P(A_2|K, A_1) = P(A_2|K).$$

If there is coupling between the systems, as illustrated above, then $P(A_2 \mid K, A_1)$ will be different from $P(A_2 \mid K)$, and thus we have a dependence between the events A_1 and A_2.

A conditional probability, $P(A|B)$, is defined by the formula

$$P(A|B) = P(A \text{ and } B)/P(B).$$

We see that this formula is simply a rewriting of formula (A.1). By substituting $P(A \text{ and } B)$ with $P(A) \; P(B|A)$ (again we use formula (A.1)), the well-known Bayes formula is established:

$$P(A|B) = P(A) \; P(B|A)/P(B).$$

We will show an application of this formula in Section A.2.2.

A.1.3 Random variables (quantities) and probability distributions

In applications we often focus on one or more summarising performance measures, in contrast to all possible outcomes. As an example, let us consider k gas detectors in a process plant. We are primarily interested in the *number* of detectors that are not functioning, i.e. not raising the alarm. Let X denote this number. The value of X is uniquely given when the outcome of the "experiment" is registered. If, for example $k = 2$ and it is observed that detector 1 is functioning but not detector 2, then $X = 1$. Thus we may view X as a function from the sample space to the real numbers. We call such variables random variables or stochastic variables. If the subjective probability interpretation is adopted, it is common to refer to X as a random quantity, or simply an unknown or uncertain quantity. The word "variable" is usually avoided as it gives the wrong impression that X varies. In the following we will use the term "random quantity" as the generic term, and refer to random variables only when interpreting probability in a classical or relative frequency way.

Let X denote a random quantity and assume that X is discrete, i.e. it can only take a finite number of values or a countable infinite number of values. Let $P(X = x)$ denote the probability of the event "$X = x$", where x is one of the values X can take. We call the function $f(x) = P(X = x)$ the probability distribution of X, or simply the distribution of X. The cumulative probability distribution $F(x)$ is defined by $F(x) = P(X \leq x)$. It is also referred to as the probability distribution of X.

In many applications we prefer to work with random quantities having continuous distributions, i.e. distributions which are characterised by a probability density $f(x)$ such that

$$P(a < X \leq b) = \int_{(a,b]} f(x) \; dx.$$

Thus if $b - a$ is small, $P(a < X \leq b) \approx f(x)(b - a)$.

The probability distribution $F(x)$ is defined by $F(x) = P(X \leq x)$.

Expected value and variance

Let X denote the number of failed systems in the course of one year for a group of four systems. Assume that you have established the following probabilities that X will take the value i, i = 0, 1, 2, 3, 4:

i	0	1	2	3	4
P(X = i)	0.05	0.40	0.40	0.10	0.05

The expectation EX is defined by:

$$EX = 0 \cdot 0.05 + 1 \cdot 0.40 + 2 \cdot 0.40 + 3 \cdot 0.10 + 4 \cdot 0.05 = 1.7.$$

The expected value is the centre of gravity of the distribution of X. If a lever is set up over the point 1.7, then the masses 0.05, 0.40, ..., 0.05 over the points 0, 1, ..., 4 will be perfectly balanced.

If X can assume one of the following values, $x_1, x_2, \ldots$, one can find the EX value by multiplying x_1 with the corresponding probability P_1, and likewise multiply value x_2 with probability P_2, etc., and sum up all values x_i, i.e.

$$EX = x_1 \cdot P_1 + x_2 \cdot P_2 + \cdots$$

If X indicates the number of events of a specific type, and this number is either zero or one, the associated probability $P(X = 1)$ equals the expected value. This is evident from the formula for expected value as in this case EX is equal to $1 \cdot P$(the event will occur). In many situations, we are concerned about rare events in which we, for all practical purposes, can disregard the possibility of two or more of such events occurring during the time interval under consideration. The expected number of events will then be approximately equal to the probability that the event will occur once.

In applications, we often use the term "frequency" for the expected value with respect to the number of events. We speak of the frequency of gas leakages, for example, and we actually mean the expected value. We can also regard the frequency as an observation, or prediction, of the number of events during the course of a specific period of time. If we, for example, say that the frequency is two per year, we have observed, or we predict, two events per year on average.

The expectation constitutes the centre of gravity of the distribution, as mentioned above, and we see from the example distribution that the actual outcome can be far from the expected value. To describe the uncertainties, a prediction interval is often used. A 90% prediction interval for X is an interval [a, b], where a and b are constants, which are such that

$P(a \leq X \leq b) = 0.90$. In cases where the probabilities cannot be determined such that the interval has probability 0.90, the interval boundaries are specified such that the probability is larger than, and as close as possible to, 0.90. In our example, we see that [1, 3] is a 90% prediction interval. We are 90% certain that X will assume one of the values 1, 2 or 3.

The variance and standard deviation are used to express the spread around the expected value. The variance of X, Var X, is defined as the expectation of $(X - EX)^2$, i.e. $\text{Var} X = E[(X - EX)^2]$, while the standard deviation is defined as the square root of the variance.

Independence

Let $X_1, X_2, \ldots, X_n$ denote n arbitrary random quantities. We say that these quantities are independent if

$$P(X_1 \leq x_1, X_2 \leq x_2, \ldots, X_n \leq x_n) = P(X_1 \leq x_1) \, P(X_2 \leq x_2) \cdot \cdots \cdot P(X_n \leq x_n)$$

for all choice of $x_1, x_2, \ldots, x_n$. In a subjective probability context independence means judged independence.

Exchangeability

Consider two discrete random quantities X_1 and X_2. Then X_1 and X_2 are said to be exchangeable if for all values x_1 and x_2 that X_1 and X_2 can take, we have

$$P(X_1 = x_1 \text{ and } X_2 = x_2) = P(X_1 = x_2 \text{ and } X_2 = x_1),$$

that is, the assessed probabilities are unchanged (invariant) by switching (permuting) the indices.

More generally, random quantities $X_1, X_2, \ldots, X_n$ are exchangeable if their joint distribution is invariant under permutations of coordinates, i.e.

$$F(x_1, x_2, \ldots, x_n) = F(x_{r1}, x_{r2}, \ldots, x_{rn}),$$

where F is a generic joint cumulative distribution for $X_1, X_2, \ldots, X_n$ and equality holds for all permutation vectors (r1,r2, ...,rn).

Exchangeability is weaker than independence, because in general exchangeable random quantities are dependent. Independent random quantities having identical probability distributions are exchangeable (but not vice versa). In a subjective probability context, exchangeability means judged exchangeability.

The strong law of large numbers

The following theorem, known as the strong law of large numbers, is one of the most well-known results in probability theory. It states that the average of a sequence of independent random quantities having the same distribution will, with probability 1, converge to the mean of that distribution.

Let X_1, X_2, ... be a sequence of independent random quantities having a common distribution, and let $EX_i = \mu$. Then with probability one,

$$(X_1 + X_2 + \cdots + X_n)/n \to \mu$$

$$\text{as } n \to \infty.$$

Binomial and Poisson distributions

Let us imagine that we have a large population, I, of similar systems, and we are studying the proportion q of them that will fail over the course of the next year. Let us imagine further that we have another similar population II that is composed of n systems. Let X represent the number that fail in this population. What then is our probability that all of those in population II will fail, i.e. $P(X = n)$?

To answer this question, first assume that q is known. You know that the proportion q within the larger population I is 0.10 say. Then the problem boils down to determining $P(X = n \mid q)$. If we do not have any other information it would be natural to say that

$$P(X = n|q) = q^n.$$

We have n independent trials and our probability for "success" (failure) is q in each of these trials. We see that when q is known, then X has a so-called binomial probability distribution, i.e.

$$P(X = i|q) = n!/\left[i!(n-i)!)\right] q^i(1-q)^{n-i} \; i = 0, 1, 2, \ldots n,$$

where $i! = 1 \cdot 2 \cdot 3 \ldots \cdot i$. The reader is referred to a textbook on probability calculus if understanding this is difficult. The mean and variance of the binomial distribution are equal to nq and $nq(1-q)$, respectively.

When q is small and n is large, we can approximate the binomial probability distribution by using the Poisson distribution:

$$P(X = i|r) = r^i \, e^{-r}/i!, i = 0, 1, 2, \ldots, \tag{A.2}$$

where $r = nq$. We know, for example, that $(1-q)^n$ is approximated equal to e^{-r}. Check this using a pocket calculator.

We refer to q and r as parameters in the probability distributions. By varying the parameters, we obtain a class of distributions. The mean and variance in a Poisson distribution are both equal to the parameter r.

What do we do if q is unknown? Let us imagine that q can be 0.1, 0.2, 0.3, 0.4, or 0.5. We then use the law of total probability to obtain that

$$P(X = i) = P(X = i|q = 0.1) \, P(q = 0.1) + P(X = i|q = 0.2) \, P(q = 0.2) + \cdots +$$
$$P(X = i|q = 0.5) \, P(q = 0.5).$$

By setting values for $P(q = 0.1)$, $P(q = 0.2)$, etc., we obtain the probability distribution for X, i.e. $P(X = i)$ for various values of i.

Uniform distribution

A random quantity X is uniformly distributed on the interval [a, b] if it has a probability density given by

$f(x) = 1/(b - a)$, for $a < x < b$, and $f(x) = 0$ otherwise. The mean and variance of X are equal to $(b - a)/2$ and $(b - a)^2/12$, respectively.

Exponential distribution

A random quantity X is said to be exponentially distributed with parameter λ if

$$P(X \le x) = 1 - e^{-\lambda x}, \text{for } x > 0.$$

Often an exponential lifetime distribution is used for describing the lifetime of a unit. For this distribution we have $P(X>u+v \mid X>u) = P(X>v)$, which means that the probability of survival of the additional v units of time is not dependent on how long the unit has functioned. The exponential distribution is the only distribution with this property. This lack of memory simplifies the mathematical analysis.

An important quantity in studying lifetime distributions is the so-called failure rate $z(x)$, defined by

$$z(x) = f(x)/(1 - F(x)), \tag{A.3}$$

where $F(x) = P(X \le x)$ and $f(x)$ is the corresponding probability density function. For the exponential distribution, the failure rate is equal to λ, i.e. independent of time. To see the physical interpretation of the failure rate, consider a small time interval (x, x+h) and assume that the unit has survived x. Then we find that

$$(1/h) \, P(X \le x + h|X>x) = (1/h) \, P(x<X \le x + h)/P(X>x) =$$
$$[F(x + h) - F(x)]/h \cdot 1/(1 - F(x)) \to f(x)/(1 - F(x)) = z(x) \text{ as } h \to 0.$$

Thus

$$P(X \le x + h|X>x) \approx z(x) \, h$$

for small values of h. We see that the failure rate expresses the proneness of the unit to fail at time (age) x. A high failure rate means that there is a high probability that the unit will fail "soon", whereas a small failure rate means that there is a small probability that the unit will fail in a short time. The cumulative failure rate $\int_{[0,x]} z(t)dt$ is known as the hazard and is denoted $Z(x)$.

The mean and variance in the exponential distribution are given by

$$EX = 1/\lambda \text{ and } \operatorname{Var} X = 1/\lambda^2.$$

Weibull distribution

A random quantity X is said to be Weibull distributed with parameters λ and β if the distribution is given by

$$P(X \leq x) = 1 - \exp\left\{-(\lambda x)^\beta\right\} \text{ for } x>0.$$

We call λ and β the scale and form parameter, respectively. If $\beta = 1$, the failure rate becomes a constant. Hence the exponential distribution is a special case of the Weibull distribution. When $\beta > 1$, the failure rate is increasing, and when $\beta < 1$, it is decreasing. Note that

$$1 - F(1/\lambda) = \exp\{-1\} = 0.3679.$$

The quantity $1/\lambda$ is often called the characteristic lifetime. The mean (expected) lifetime of the Weibull distribution is given by

$$EX = (1/\lambda)\Gamma(1 + 1/\beta),$$

where Γ is the gamma function defined by

$$\Gamma(x) = \int_{[0,\infty)} t^{x-1}e^{-t}dt.$$

In particular, we have $\Gamma(n+1) = n!$ for $n = 0, 1, 2, \ldots$

The variance of X equals

$$\operatorname{Var} X = (1/\lambda^2)[\Gamma(1 + 2/\beta) - \Gamma^2(1 + 1/\beta)].$$

Gamma distribution

If $X_1, X_2, \ldots, X_n$ are independent and exponentially distributed random quantities with parameter λ, then $X_1 = X_2 + \cdots + X_n$ is gamma distributed with parameters λ and n, i.e.

$$f(x) = \lambda(\lambda x)^{n-1}e^{-\lambda x}/\Gamma(n) \quad x > 0. \tag{A.4}$$

Assume for example that n units of a certain type have exponentially distributed lifetimes $X_1, X_2, \ldots, X_n$ with failure rate λ and that the units are put into operation one by one as a unit fails. Then the total lifetime equals the sum of the random quantities X_i.

The parameter n in (A.4) does not need to be restricted to the positive integers. If it is a positive integer, we can write the survivor function in the following form:

$$1 - F(x) = \Sigma_{\{i=0, \ldots, n-1\}} (\lambda x)^i \, e^{-\lambda x}/i!.$$

The mean and variance of the gamma distribution are given by

$$EX = n/\lambda$$
$$Var\,X = n/\lambda^2.$$

Chi-square distribution

A random quantity X is chi-square distributed with parameter v if it has a density given by

$$f(x) = x^{(v/2)-1} e^{-x/2}/\{2^{v/2}\Gamma(v/2)\} \quad x > 0.$$

The mean of the distribution equals v and the variance 2v. The chi-square distribution is closely linked to the Gamma distribution. If X has a Gamma distribution with parameters (n, λ), then $2\lambda X$ is chi-square distributed with parameter 2n.

Beta distribution

A random quantity X is said to be beta distributed with parameters a and b if it has a density given by

$$f(x) = [\Gamma(a+b)/(\Gamma(a)\,\Gamma(b))]x^{a-1}(1-x)^{b-1}, \text{for } x>0, a>0, b>0.$$

The mean and variance are equal to $a/(a+b)$ and $ab/[(a+b)^2(a+b+1)]$, respectively.

Beta-binomial distribution

A random quantity X is said to be beta-binomial distributed with parameters (n,a,b) if it has a density given by

$$f(x) = (n!/(n-x)!\,x!)[(\Gamma(a+x)\,\Gamma(n+b-x)\Gamma(a+b)/\{\Gamma(n+a+b)\,\Gamma(a)\,\Gamma(b)\}],$$

for $x = 0, 1, 2, \ldots, n$, $a > 0$, $b > 0$ and $n = 0, 1, 2, \ldots$. The mean and variance are equal to $na/(a+b)$ and $nb(n+a+b)/[(a+b)^2(a+b+1)]$, respectively.

Triangular distribution

A random quantity X is triangle distributed with parameters a,b and c if it has a density given by

$$f(x) = 2(x-a)/[(b-a)(c-a)] \quad \text{if } a \leq x \leq b, \text{ and}$$
$$f(x) = 2(c-x)/[(c-a)(c-b)] \quad \text{if } b < x \leq c.$$

The density increases linearly from a to b, and then decreases linearly from b to c. The mean and variance are equal to $(a+b+c)/3$ and $(a^2 + b^2 + c^2 - ab - ac - bc)/18$, respectively.

Normal distribution

The random quantity X is said to be normally distributed with parameters μ and σ if it has a density given by

$$f(x) = (1/2\pi)^{1/2} \exp\left\{-((x - \mu)/\sigma)^2\right\}.$$

It can be shown that $EX = \mu$ and $\text{Var } X = \sigma^2$. If $\mu = 0$ and $\sigma = 1$, the distribution is referred to as a standard normal distribution. The normal distribution is probably the most widely used distribution in the entire field of statistics and probability. It turns out that the mean of a number of quantities X is normally distributed. The central limit theorem gives a precise mathematical formulation of this fact:

Central limit theorem

Let X_1, X_2, ... be a sequence of independent random quantities having a common distribution, and let $EX_i = \mu$ and $\text{Var } X_i = \sigma^2$. Then the distribution of

$$n^{1/2}[(X_1 + X_2 + \cdots + X_n)/n - \mu]/\sigma$$

converges to the standard normal distribution with mean 0 and variance 1.

Student (t) distribution

A random quantity X is Student (t) distributed with parameter v (degrees of freedom) if it has a density given by

$$f(x) = (1 + x^2/v)^{-(v+1)/2} \Gamma((v + 1)/2)/\{\Gamma(v/2)(\pi v)^{1/2}\}.$$

If W and Z are independent, W with a standard normal distribution and Z has a chi-square distribution with parameter v, then $T = W/(Z/v)^{1/2}$ has a Student (t) distribution with parameter v.

A.2 Statistical analysis

A.2.1 Traditional statistical analysis

Non-parametric estimation

Consider a random variable X, having probability distribution $F(x) = P(X \leq x)$. The task is to estimate this distribution given observations $X_1, X_2, \ldots, X_n$. The random variable X_i has a distribution function F. All the random variables are assumed independent.

Often the data are censored, i.e. we do not observe X_i, but minimum$\{X_i, C_i\}$, where C_i is the censoring time. We will, however, not discuss this case any further here.

As an estimator for F(x) we may use the empirical distribution function $F^*(x)$ defined by

$$F^*(x) = (1/n) \, \Sigma I(X_i \leq x),$$

where I is the indicator function which equals 1 if the argument is true and 0 otherwise. If $n \to \infty$, then $F^*(x) \to F(x)$ with probability one.

For non-negative observations the Nelson–Aalen estimator Z^* is often used. This is an estimator of the cumulative failure rate $Z(x) = \int_{[0,x]} z(t)dt$, cf. (A.3), and is given by

$$Z^*(t) = \Sigma_{\{i:,x_i \leq t\}} 1/(n - i + 1).$$

Based on estimators as above we can make plots and fit the distribution to a parametric class of distributions, like the exponential distribution.

If we compute the Nelson–Aalen estimator and the plot is close to a straight line starting in origo, this would indicate that an exponential distribution may be appropriate as the hazard of this distribution is such a straight line.

In this framework, we may also use so-called "goodness of fit" tests. The idea is to use a measure of distance between the empirical distribution and the underlying theoretical distribution. We refer to textbooks in statistics.

Estimation of distribution parameters

We assume that the distribution F(x) belongs to a known parametric class of distributions, for example the exponential or the normal distribution. The problem is to estimate the parameters of the distribution. As above we assume that we have observations $X_1, X_2, \ldots, X_n$.

Maximum likelihood estimation   As a first illustrating example, we consider the Poisson distribution. Let

$$f(x|\lambda) = \lambda^x e^{-\lambda}/x!.$$

The probability density related to the observed data $X_i = x_i$ then becomes

$$f(x_1|\lambda)f(x_2|\lambda) \cdot \ldots \cdot f(x_n|\lambda) = c\, \lambda^{x_1+x_2+\cdots+x_1} e^{-\lambda},$$

where c is a quantity that does not depend on λ. As a function of the parameter λ, this expression is called the likelihood function and is denoted $L(\lambda)$. The likelihood function is not a probability distribution. It is interpreted as a scale of comparative support lent by the known x_i values to the various possible values of the unknown λ (Singpurwalla 2006).

The maximum likelihood estimate (MLE) of λ, λ^*, maximises $L(\lambda)$. In other words, MLE is the value of λ that makes the observed result most likely. In practice the MLE is determined by differentiating the likelihood function and setting the derivative equal to zero. By doing so, we obtain for the example

$$\lambda^* = x_1 + x_2 + \cdots + x_n/n,$$

i.e. the average number of the observations.

Confidence interval

As a measure of data variation, a confidence interval for the parameter is often presented in addition to the estimate of the parameter. As an example consider the binomial distribution with parameters (n,p) with n known and p unknown. It is assumed that n is large (>30). The MLE estimator of p is p^* given by $p^* = X/n$, i.e. the success rate. The expected value and variance of this estimator are equal to

$$Ep^* = p$$
$$Var(p^*) = p(1-p)/n.$$

By the central limit theorem, p^* has an approximate normal distribution as X is the sum of n independent random quantities which are either zero or one. Hence $(p^* - p)/[p(1-p)/n]^{1/2}$ has an approximate standard normal distribution, and as p^* is close to p by the strong law of large numbers, also $(p^* - p)/[p^*(1-p^*)/n]^{1/2}$ has an approximate standard normal distribution. Let d denote the estimated standard deviation term $[p^*(1-p^*)/n]^{1/2}$.

Then by using statistical tables for the standard normal distribution we obtain

$$P(-1.645 \leq (p^* - p)/d \leq 1.645) \approx 0.90, \text{ i.e.}$$
$$P(p^* - 1.645d \leq p \leq p^* + 1.645d) \approx 0.90.$$

The interval $[p^* - 1.645d, p^* + 1.645d]$ is an approximate 90% confidence interval for p. There is a probability of about 90% that this interval contains the underlying correct p value. Note that when the confidence interval is calculated, i.e. we observe X, the resulting interval either contains the true value of p or it does not, but in the long run if the experiment were repeated many times, the confidence interval would include p in about 90% of the times.

In the exponential model with parameter λ, and observations x_i, $i = 1, 2, \ldots, n$, a $(1 - \alpha)100\%$ confidence interval is given by

$$(\lambda_L, \lambda_H) = (z_1/2y, \ z_2/2y), \tag{A.5}$$

where y is the sum of the n observations, z_1 equals the $\alpha/2$ quantile in the chi-square distribution with 2n degrees of freedom and z_2 equals the $1 - \alpha/2$ quantile in the chi-square distribution with 2n degrees of freedom. The α quantile of the distribution of a random variable X is the value x such that $P(X \leq x) = \alpha$.

For the Poisson process model with rate λ and observed in an interval t, we can use the same interval (A.5) with z_1 equal to the $\alpha/2$ quantile in the chi-square distribution with 2N degrees of freedom and z_2 equal to the $1 - \alpha/2$ quantile in the chi-square distribution with $2(N+1)$ degrees of freedom, where N is the number of events observed in the interval [0, t].

Testing hypothesis

The set-up is as above. We assume that the distribution F(x) belongs to a known parametric class of distributions and that we have available observations $X_1, X_2, \ldots, X_n$. We use the binomial model with parameters n and p to illustrate ideas. The observation X_i here refer to "success" in the ith experiment, such that the sum of the X_is is the total number of observed "successes". This sum is prior observation seen as a random variable, and we denote it by Y.

The problem is now to formulate a statistical test. We do this by formulating statements about the parameter of the probability model, in this case the success probability p. The starting point is the null-hypothesis H_0, which we may think of as "p = 0.25", say. The test questions the truth of this statement in relation to an alternative hypothesis H_1, say p > 0.25. If the data provide sufficient support, we assert that H_0 is false and H_1 is correct. We conclude in this way if we have a high confidence about the correctness of H_1. As a concrete example, consider a medical treatment that is known to have a "success" rate of 25 per cent. An adjustment of this treatment is considered, and the question is whether this adjustment would increase the "success" rate.

It is reasonable to assert that $p > 0.25$ if the number of successes is large, i.e. $Y \geq k$, for a suitable choice of k. Let α be the probability that $Y \geq k$ if H_0 is true, i.e. $p = 0.25$. These probabilities for various k are found from statistical tables for the binomial distribution, or use of approximations to the normal distribution. We search for a k such that α becomes rather small, say 0.05 or 0.10. For example, if $n = 20$ and $\alpha = 0.10$, we find that $k = 8$, which corresponds to a fraction of successes of 40 per cent. If we observe eight or more successes, the result is so "extreme" relative to H_0, that we reject H_0. We refer to α as the significance level of the test. It is the probability of an error of type I, i.e. of rejecting H_0 when, in fact, it is true. It should be rather low as it represents a probability of making a wrong conclusion: asserting H_1 if H_0 is true. On the other hand, specifying a very low value of α means that the probability of not concluding that H_1 is true if it is, in fact, true, becomes high. So a balance has to be achieved. The probability of this latter type of error is denoted β, and is a function of the parameter value. This type of error is referred to as an error of type II. In our example, if $p = 0.30$, the probability that we do not reject H_0, the type II error probability $P(Y < 8 | p{=}0.3)$ is about 77 per cent. We see that to reject H_0 a rather extreme observation is required using the above principles. The point is that type I errors are considered more serious than type II errors. In the medical treatment example, the starting point is that there is no improvement. Only if the data give a very strong support for the alternative hypothesis, should we reject H_0; the probability of a failure of type I should be small. Note that when not rejecting H_0, we do not say that H_0 is true; the conclusion is that we do not have statistical evidence to reject the null hypothesis.

Regression analysis

Regression analysis is mainly used for the purpose of prediction. By developing a statistical model, the values of a dependent or response variable Y are predicted based upon the values of an independent variable X. As an example, an economist might want to develop a statistical model that predicts how much money a population of people would spend (Y) based on how much money they earn (X). The simplest type of regression analysis is based on a linear regression model. To develop the model, we assume that a sample of n independent observations (X_1, Y_1), (X_2, Y_2), ..., (X_n, Y_n) is obtained, where X_i represents the ith value of the independent variable X and where Y_i represents the corresponding response – that is, the ith value of the dependent variable Y. The linear regression model specifies that there is an underlying true relationship between EY and EX, expressed by a linear function. In practice this linear function is not realised because of randomness. Mathematically, these ideas are formulated as

$$Y_i = \alpha + \beta X_i + \varepsilon_i,$$

where ε_i is the random error in Y for observation i, and α and β are parameters to be estimated. We see that β represents the slope of the line $Y = \alpha + \beta X$, and α represents the intercept of the line with the Y axis. We may think of this underlying straight line as a model of the true relationship between EY and EX for a large (infinite) population to which the sample of n belongs. The random variables ε_i represent the error terms. A common model for these error terms is the normal distribution with mean zero and variance σ^2. This distribution reflects the variations of the observations Y around their expected values.

Now, to estimate the parameters α and β the standard technique is to apply the method of least squares, i.e. to identify the values that minimise the sum of squared errors in the sample. The estimators then derived, we denote by α^* and β^*, and they are

$$\alpha^* = \overline{Y} - \beta^*,$$
$$\beta^* = \Sigma_i(Y_i - \overline{Y})(X_i - \overline{X})/\Sigma_i(X_i - \overline{X})^2,$$

where $\overline{X}$ and $\overline{Y}$ are the means of the X_i and Y_i, respectively, $i = 1, 2, \ldots, n$. To predict Y based on X we use the line

$$Y = \alpha^* + \beta^* X.$$

To estimate the variance σ^2, the common estimator is

$$S^2 = \Sigma_i(Y_i - \alpha^* - \beta^* X_i)^2/(n - 2).$$

Confidence intervals and statistical tests can now be derived for the parameters α, β and σ^2. The slope of the line, β, is of special interest as it is a measure of the trend of the data. If $\beta = 0$ there is no trend and often the analysis is concerned about the extent to which the data prove that there is a trend present. Could the observed increase in the slope just be a result of "randomness"? To perform this analysis we need to assume a specific probability distribution for the error term ε_i, and the common choice is the normal distribution with mean 0 and variance σ^2. Then Y_i is also normally distributed, with mean $\alpha + \beta X_i$ and variance σ^2. A 90% confidence interval for β is then given by

$$\beta^* + / - t_{n-2} \cdot S_\beta,$$

where t_{n-2} is the 95% quantile of the Student (t) distribution with $n - 2$ degrees of freedom, and S_β is an estimator of the standard deviation of β^* given by

$$(S_\beta)^2 = \left[\Sigma_i(Y_i - \alpha^* - i\beta^*)^2/(n-2)\right]/\Sigma_i(X_i - \overline{X})^2.$$

Similarly, a confidence interval for $\mu_i = \alpha + \beta X_i$ can be formulated. It takes the following form:

$$\alpha^* + X_i\beta^* + / - t_{n-2} \cdot \left[\Sigma_j(Y_j - \alpha^* - X_j\beta^*)^2/(n-2)\right]^{1/2}$$
$$(1/n) + (X_i - \overline{X})/\Sigma_j(X_i - \overline{X})^2\}^{1/2}.$$

We refer to textbooks in statistics, such as Berenson *et al.* (1988).

A.2.2 Bayesian statistics

Consider the proportion q of failed systems in population I, introduced in Section A.1.3. The problem considered now is how to express our knowledge of q based on the available data X, i.e. to establish a probability distribution for q when we observe X, in a context where all probabilities are knowledge-based (subjective). We call this distribution the posterior distribution of q.

We begin with the so-called a priori distribution, before we perform the measurements X. Let us assume that we only allow q to assume one of the following five values; 0.1, 0.2, 0.3, 0.4 or 0.5. We understand these values such that, for the example 0.5, this means that q lies in the interval [0.45, 0.55).

Based on the available knowledge, we assign a prior probability distribution of the proportion q:

q'	0.1	0.2	0.3	0.4	0.5
P(q=q')	0.05	0.20	0.50	0.20	0.05

This means that we have the greatest confidence that the proportion q is 0.3 (50%), then 0.2 and 0.4 (20% each) and least likely 0.1 and 0.5 (5% each).

Suppose now that we observe 10 systems and that among these systems there is only one that has failed. How will we then express our uncertainty regarding q?

We then use Bayes' formula and establish the posteriori distribution of q. The Bayes' formula states that $P(A|B) = P(B|A)P(A)/P(B)$ for the events A and B. If we apply this formula, we see that the probability that the proportion will be equal to q' when we have observed that one out of 10 has failed, is given by:

$$P(q = q'|X = 1) = cf(1|q')P(q = q'), \tag{A.6}$$

where c is a constant such that the sum over different values of q' is equal to 1, and f is given by

$$f(i|q') = P(X = i|q = q').$$

Here X is binomially distributed with parameters 10 and q' when $q = q'$ is given.

By formula (A.6) we find the following updated posterior distribution of q:

q'	0.1	0.2	0.3	0.4	0.5
$P(q=q')$	0.14	0.38	0.43	0.05	0.004

We see that the probability mass has shifted to the left towards smaller values. This was as expected since we observed that only one out of 10 systems has failed, while we, at the start, expected the proportion q to be closer to 30%.

If we had had a larger observation set, then this data set would have dominated the distribution to an even larger degree.

Appendix B

Terminology

This appendix summarises some risk analysis and management terminology used in the book. Unless stated otherwise, the terminology is in line with the international guideline ISO (2009a).

- *aleatory (stochastic) uncertainty*: variation of quantities in a population. This definition is not given in the ISO guideline.
- *epistemic uncertainty*: lack of knowledge about unknown quantities. This definition is not given in the ISO guideline.
- *event*: occurrence or change of a particular set of circumstances.
- *frequency*: number of events per unit of time or another reference. Often frequency is also used for the expected number of events per unit of time.
- *managerial review and judgement*: process of summarising, interpreting and deliberating over the results of risk assessments and other assessments, as well as of other relevant issues (not covered by the assessments), in order to make a decision. This definition is not given in the ISO guideline.
- *probability*: either a knowledge-based (subjective) measure of uncertainty of an event conditional on the background knowledge or a relative frequency (chance). If a knowledge-based probability is equal to 0.10, it means that the uncertainty (degree of belief) is the same as randomly drawing a specific ball out of an urn. A relative frequency-interpreted probability (chance) is the fraction of events A occurring when the situation considered can be repeated over and over again infinitely. This definition is not given in the ISO guideline.
- *risk*: the book refers to different definitions. The most general and the one recommended says that risk is the two-dimensional combination of
  (i) events A and associated consequences C

(ii) uncertainties about A and C (will A occur, what will the consequences C be?).

This definition is referred to as the (A,C,U) definition.
This definition is not given in the ISO guideline.

- *risk acceptance*: a decision to accept risk.
  This definition represents an adjustment of the definition used by the ISO guideline: informed decision to take a particular risk.
- *risk acceptance criterion*: a reference by which risk is assessed to be acceptable or unacceptable.
  This definition is not included in the ISO guideline. It is an example of a risk criterion.
- *risk analysis*: systematic use of information to identify risk sources, causes and consequences of these sources, and describe risk.
  The ISO guideline does not include source identification as a part of risk analysis. It states that a risk analysis is the process to comprehend the nature of risk and to determine the level of risk.
- *risk appetite*: amount and type of risk an organisation is prepared to pursue or retain.
- *risk assessment*: the overall process of risk analysis and risk evaluation.
- *risk communication*: exchange or sharing of risk-related information between stakeholders.
  This definition is not given in the ISO (2009a) guideline, but in the ISO (2002) guideline.
- *risk criteria*: terms of reference against which the significance of the risk is evaluated.
- *risk description*: a qualitative and/or quantitative picture of the risk.
  This definition represents an adjustment of the one given in the ISO guideline.
- *risk evaluation*: process of comparing the result of risk analysis against risk criteria to determine the significance of the risk.
  This definition represents an adjustment of the definition used by the ISO guideline.
  See also managerial review and judgement.
- *risk management*: coordinated activities to direct and control an organisation with respect to risk.
- *risk perception*: stakeholder's subjective judgement or appraisal of risk.
  This definition represents an adjustment of the definition used by the ISO guideline.

- *risk quantification*: process used to assign values to the probabilities and risk indices used.
  This definition is not given in the ISO guideline.
- *risk source*: element which alone or in combination has the intrinsic potential to give rise to an event with a consequence.
  This definition represents an adjustment of the definition used by the ISO guideline.
- *risk tolerability level*: level of risk which an organisation will tolerate.
  This definition is not given in the ISO guideline.
- *risk retention*: acceptance of the potential benefit of gain, or burden of loss, from the risk.
- *risk treatment*: process to modify risk.
- *stakeholder*: person or organisation that can affect, be affected by, or perceive themselves to be affected by a decision or activity.
- *uncertainty*: lack of knowledge about unknown quantities.
  This definition is not included in the ISO guideline.
- *vulnerability*: In line with the (A,C,U) risk definition vulnerability is the two-dimensional combination of
  (i) consequences C
  (ii) uncertainties about C (what values will C take)
  given the occurrence of A. We write (C,U|A).
  This definition is not included in the ISO guideline.

References

Abrahamsen, E. B. and Aven, T. (2011). Safety oriented bubble diagrams in project risk management. *International Journal of Performability Engineering*, **7**(1), 91–96.

Abrahamsen, E. B., Aven, T. and Røed, W. (2010). Communication of cost-effectiveness of safety measures by use of a new visualizing g tool, *Reliability & Risk Analysis: Theory & Applications*, **2**(4), 38–46.

Abramson, L. R. (1981). Some misconceptions about the foundations of risk analysis. *Risk Analysis*, **1**, 229–230.

Ale, B. J. M. (2002). Risk assessment practices in The Netherlands. *Safety Science*, **40**:105–126.

Ale, B., Bellamy, L. J., van der Boom, R. *et al.* (2009). Further development of a Causal model for Air Transport Safety (CATS): Building the mathematical heart. *Reliability Engineering and System Safety*, **94**, 1433–1441.

Anoop, M. B. and Rao, K. B. (2008). Determination of bounds on failure probability in the presence of hybrid uncertainties. *Sadhana*, **33**, 753–765.

Apostolakis, G. E. (1990) The concept of probability in safety assessments of technological systems. *Science*, **250**, 1359–1364.

Apostolakis, G. E. (ed.) (1988). *Reliability Engineering and System Safety*, **23**.

Apostolakis, G. E. (1990). The concept of probability in safety assessments of technological systems. *Science*, **250**, 1359–1364.

Apostolakis, G. E. (2004). How useful is quantitative risk assessment? *Risk Analysis*, **24**, 515–520.

Apostolakis, G. E. and Pickett, S. E. (1998). Deliberation: Integrating analytical results into environmental decisions involving multiple stakeholders, *Risk Analysis*, **18**(5), 621–634.

AS/NZS 4360 (2004). Australian/New Zealand Standard: *Risk management*.

Aven, T. (1986) Formulae for the average unavailability (MFDT) of a coherent system with periodically tested components. *Microelectronics and Reliability*, **26**, 283–288.

Aven, T. (1992). *Reliability and Risk Analysis*. London: Elsevier.

Aven, T. (2003). *Foundations of Risk Analysis*. New Jersey: Wiley.

Aven, T. (2004). Risk analysis and science. *International Journal of Reliability, Quality and Safety Engineering*, **11**, 1–15.

Aven, T. (2006). On the precautionary principle, in the context of different perspectives on risk. *Risk Management: an International Journal*, **8**, 192–205.

Aven, T. (2007a). A unified framework for risk and vulnerability analysis and management covering both safety and security. *Reliability Engineering and System Safety*, **92**, 745–754.

Aven, T. (2007b). On the ethical justification for the use of risk acceptance criteria. *Risk Analysis*, **27**, 303–312.

Aven, T. (2008a). *Risk Analysis*, New Jersey: Wiley.

Aven, T. (2008b). A semi-quantitative approach to risk analysis, as an alternative to QRAs. *Reliability Engineering and System Safety*, **93**, 768–775.

Aven, T. (2009a). Perspectives on risk in a decision-making context – Review and discussion. *Safety Science*, **47**, 798–806.

Aven, T. (2009b). Safety is the antonym of risk for some perspectives of risk. *Safety Science*, **47**, 925–930.

Aven, T. (2009c). A new scientific framework for quantitative risk assessments *International Journal of Business Continuity and Risk Management*, **1**(1), 67–77.

Aven, T. (2010a). *Misconceptions of Risk*, Chichester: Wiley.

Aven, T. (2010b). Some reflections on uncertainty analysis and management. *Reliability Engineering and System Safety*, **95**, 195–201.

Aven, T. (2010c). On the need for restricting the probabilistic analysis in risk assessments to variability. *Risk Analysis*, **30**(3), 354–360.

Aven, T. (2010d). Reply to discussants on "The need for restricting the probabilistic analysis in risk assessments to variability". *Risk Analysis*, **30** (3), 381–384.

Aven, T. (2010e). On how to define, understand and describe risk. *Reliability Engineering and System Safety*. **95**, 623–631.

Aven, T. (2010f). A holistic framework for conceptualising and describing risk. In Proceedings SSARS conference, Gdansk 20–25 June, 2010.

Aven, T. (2010g). On different types of uncertainties in the context of the precautionary principle. Revised and resubmitted *Risk Analysis*.

Aven, T. (2010h). Selective critique of risk assessments with recommendations for improving methodology and practice. Revised and resubmitted *Reliability Engineering and System Safety*.

Aven, T. (2010i). Shaky foundations: common misconceptions in risk assessment and management, and ideas for fixing them. Revised and resubmitted *Risk Analysis*.

Aven, T. (2011). On some recent definitions and analysis frameworks for risk, vulnerability and resilience. *Risk Analysis*. To appear.

Aven, T. and Abrahamsen, E. B. (2007). On the use of cost-benefit analysis in ALARP processes. *International Journal of Performability Engineering*, **3**, 345–353.

Aven, T., Asche, F., Lindøe, P., Toft, A., Wiencke, H. S. (2010). A framework for decision support on HSE regulations, Como, Italy: SRA Europe. 2005. In *Risks Challenging Publics, Scientists and Governments*, ed. S. Menoni. London: CRC Press, pp. 49–56.

Aven, T. and Flage, R. (2009). Use of decision criteria based on expected values to support decision-making in a production assurance and safety setting. *Reliability Engineering and System Safety*. **94**, 1491–1498.

Aven, T. and Guikema, S. (2010) Whose uncertainty assessments (probability distributions) does a risk assessment report: the analysts' or the experts'? Paper revised and resubmitted to *Reliability Engineering and System Safety*.

Aven, T. and Heide, B. (2009). Reliability and validity of risk analysis. *Reliability Engineering and System Safety*, **94**, 1862–1868.

Aven, T. and Jensen, U. (1999). *Stochastic Models in Reliability*, New York: Springer.

Aven, T. and Nøkland, T. E. (2010). On the use of uncertainty importance measures in reliability and risk analysis. *Reliability Engineering and System Safety*, **95**, 127–133.

Aven, T. and Renn, O. (2009a). On risk defined as an event where the outcome is uncertain. *Journal of Risk Research*, **12**, 1–11.

Aven, T. and Renn, O. (2009b). The role of quantitative risk assessments for characterizing risk and uncertainty and delineating appropriate risk management options, with special emphasis on terrorism risk. *Risk Analysis*, **29**, 587–600.

Aven, T. and Renn, O. (2011) *Risk Management and Governance*. New York: Springer.

Aven, T., Renn, O. and Rosa, E. (2010) The ontological status of the concept of risk. Paper submitted for possible publication.

Aven, T. and Vinnem, J. E. (2007). *Risk Management, with Applications from the Offshore Oil and Gas Industry*, New York: Springer.

Aven, T., Vinnem, J. E. and Røed, W. (2006). On the use of goals, quantitative criteria and requirements in safety management. *Risk Management: an International Journal*, **8**, 118–132.

Aven, T. and Zio, E. (2011). Treatment of uncertainties in risk assessment for practical decision-making. Accepted for publication in *Reliability Engineering and System Safety*, **96**, 64–74.

Baraldi, P. and Zio, E. (2008) A combined Monte Carlo and possibilistic approach to uncertainty propagation in event tree analysis. *Risk Analysis*, **28** (5), 1309–1325.

Bedford, T. and Cooke, R. (2001). *Probabilistic Risk Analysis*, Cambridge: Cambridge University Press.

Berenson, M. L., Levine, D. M. and Rindskopf (1988). *Applied Statistics*, New Jersey: Prentice Hall.

Berger, J. (1994). An overview of robust Bayesian analysis. *Test*, **3**, 5–124.

Bergman, B. (2009). Conceptualistic pragmatism: a framework for Bayesian analysis? *IIE Transactions*, **41**, 86–93.

Bernardo, J. and Smith, A. (1994). *Bayesian Theory*. New York: Wiley.

Bernoulli, J. (1713) Wahrrscheinlichkeitsrechnung, third and fourth parts, Ostwald, 506. Quarterly J. of Economics. Klassiker der exakten Wissenschaften, **108**, 1896, Leipzig, translation of "ars conjectandi," published in 1713.

Cabinet Office (2002). Risk: improving government's capability to handle risk and uncertainty. Strategy unit report. UK.

Campbell, S. (2005). Determining overall risk. *Journal of Risk Research*, **8**, 569–581.

Carpi, A. and Egger, A. E. (2003) The Scientific Method. *Visionlearning* Vol. *SCI-1 (1)*, 2003. www.visionlearning.com/library/module_viewer.php?mid = 45. Accessed 3 March 2010.

Cooke, R. M. (1991). *Experts in Uncertainty: Opinion and Subjective Probability in Science*. New York: Oxford University Press.

Coolen, F. P. A. and Utkin, L. V. (2007). Imprecise reliability: A concise overview. In *Risk, Reliability and Societal Safety*, eds. T. Aven and J. E. Vinnem, Proceedings of the European Safety and Reliability Conference 2007 (ESREL 2007), Stavanger, Norway, 25–27 June 2007. London: Taylor & Francis Group, pp. 1959–1966.

Cumming, R. B. (1981). Is Risk Assessment A Science? *Risk Analysis*, **1**, 1–3.

de Finetti, B. (1974). *Theory of Probability*, New York: Wiley.

de Laplace, P. S. (1814). *Theorie analytique des probabilities*. Paris: Courcier Imprimeur.

Dempster, A. (1967). Upper and lower probabilities induced by a multivalued mapping. *Annals of Mathematical Statistics*, **38**, 325–339.

de Rocquigny, E., Devictor, N. and Tarantola, S. (eds.) (2008). *Uncertainty in Industrial Practice. A guide to quantitative uncertainty management*. New Jersey: Wiley.

Devooght, J. (1998). Model uncertainty and model inaccuracy. *Reliability Engineering and System Safety*, **59**, 171–185.

Douglas, E. J. (1983). *Managerial Economics: Theory, Practice and Problems*, 2nd edn. New Jersey: Prentice Hall.

Dubois, D. (2006). Possibility theory and statistical reasoning. *Computational Statistics and Data Analysis*, **51**, 47–69.

Dubois, D. (2010). Representation, propagation and decision issues in risk analysis under incomplete probabilistic information. *Risk Analysis*, **30**, 361–368.

Dubois, D. and Prade, H. (1988). *Possibility Theory*. New York: Plenum Press.

Dubois, D., Prade, H. and Sandri, S. (1993). On possibility/probability transformations. In *Fuzzy Logic: State of the Art*, eds. R. Lowen and M. Roubens. Dordrecht: Kluwer Academic Publishers. pp. 103–112.

Ersdal, G. and Aven, T. (2008). Risk management and its ethical basis. *Reliability Engineering and System Safety*, **93**, 197–205.

European Commission/Health and Consumer Protection Directorate General, *Directorate C* (2000). Scientific Opinions: First Report on the Harmonisation of Risk Assessment Procedures, EU, Brussels.

European Commission (2003). *Final Report on Setting the Scientific Frame for the Inclusion of New Quality of Life Concerns in the Risk Assessment Process*, EU, Brussels.

Ferson, S. and Ginzburg, L. R. (1996). Different methods are needed to propagate ignorance and variability. *Reliability Engineering and System Safety*, **54**, 133–144.

Flage, R. and Aven, T. (2009). Expressing and communicating uncertainty in relation to quantitative risk analysis (QRA). *Reliability & Risk Analysis: Theory & Applications*, **2**(13), 9–18.

Flage, R., Aven, T. and Zio, E. (2009) Alternative representations of uncertainty in reliability and risk analysis – review and discussion. In *Safety, Reliability and Risk Analysis. Theory, Methods and Applications*, eds. S. Martorell, C. Guedes Soares and J. Barnett, Proceedings of the European Safety and Reliability Conference 2008 (ESREL 2008), Valencia, Spain, 22–25 September 2008. London: CRC Press pp. 2081–2091.

Flage, R., Baraldi, P., Ameruso, F., Zio, E. & Aven, T. (2010). Handling epistemic uncertainties in fault tree analysis by probabilistic and possibilistic approaches. In *Reliability, Risk and Safety: Theory and Applications*, eds. R. Bris, C. Guedes Soares and S. Martorell Supplement Proceedings of the European Safety and Reliability Conference 2009 (ESREL 2009), Prague, Czech Republic, 7–10 September 2009.

Granger Morgan, M. and Henrion, M. (1990) *Uncertainty. A Guide to Dealing with Uncertainty in Quantitative Risk and Policy Analysis*. Cambridge: Cambridge University Press.

Garrick, B. J. (2010). Interval analysis versus probabilistic analysis. *Risk Analysis*, **3**, 369–70.

Graham, J. D. (1995). Verifiability isn't everything. *Risk Analysis*, **15**, 109.

Guikema, S. and Aven, T. (2010) Is ALARP applicable to the management of terrorist risks? *Reliability Engineering and System Safety*, **95**, 823–827.

Gärdenfors, P. and Sahlin, N -E. (1988). Unreliable probabilities, risk taking, and decision making. In *Decision, Probability, and Utility*, eds. P. Gärdenfors and N -E. Sahlin. Cambridge: Cambridge University Press, pp. 313–334.

Haimes, Y. Y. (2004). *Risk Modelling, Assessment, and Management*, 2nd edn. New Jersey: Wiley.

Hamada, M. S., Wilson, A. G., Reese, C. S. and Martz, H. F. (2008). *Bayesian Reliability*. New York: Springer.

Helton, J. C. (1994). Treatment of uncertainty in performance assessments for complex systems. *Risk Analysis*, **14**, 483–511.

Helton, J. C., Johnson, J. D., Sallaberry, C. J. and Storlie, C. B. (2006). Survey of sampling-based methods for uncertainty and sensitivity analysis. *Reliability Engineering and System Safety*, **91**, 1175–1209.

Hertz, D. B. and Thomas, H. (1983). *Risk Analysis and its Applications*. Chichester: Wiley.

HSE (2001). *Reducing Risk, Protecting People*. HSE Books, ISBN 0 71762151 0.

HSE (2006). Offshore Installations (Safety Case) Regulations 2005 regulation 12 demonstrating compliance with the relevant statutory provisions.

HSE (2000). *Offshore Hydrocarbon Release Statistics, 1999*, Offshore Technology Report OTO 079, HSE Offshore Safety Division (OSD), January 2000.

Hollnagel, E. (2004) *Barriers and Accident Prevention*, Aldershot: Ashgate.

Hollnagel, E., Woods, D. D. and Leveson, N., eds. (2006). *Resilience Engineering, Concepts and Precepts*. Burlington: Ashgate.

Holton, G. A. (2004). Defining risk. *Financial Analysis Journal*, **60**, 19–25.

House of Lords (2006). *Government Policy on the Management of Risk*. Volume 1: Report. London: The Stationery Office.

Huber, W. A. (2010). Ignorance Is Not Probability. *Risk Analysis*, **3**, 371–376.

IAEA (International Atomic Energy Agency) (1995). Guidelines for Integrated Risk Assessment and Management in Large Industrial Areas, *Technical Document: IAEA–TECDOC PGVI–CIJV*, IAEA, Vienna.

IEC (1993). Guidelines for Risk Analysis of Technological Systems, *Report IEC–CD (Sec) 381 issued by Technical Committee QMS/23*, European Community, Brussels.

IPCC Fourth Assessment Report, *Climate Change* (2007). Geneva, Switzerland. www.ipcc.ch/publications_and_data/publications_ipcc_fourth_assessment_report_synthesis_report.htm. Accessed 23 April 2010.

IPCS and WHO (World Health Organization) (2004). *Risk Assessment Terminology*, Geneva: WHO.

IRGC (International Risk Governance Council) (2005). Risk Governance – Towards an Integrative Approach, White Paper no 1, O. Renn with an Annex by P. Graham, Geneva: IRGC.

ISO (2002). Risk Management Vocabulary. *ISO/IEC Guide* **73**.

ISO (2009a). Risk Management – Vocabulary. *Guide* **73**:2009.

ISO (2009b). Risk Management – Principles and guidelines, ISO 31000:2009.

Jones-Lee, M. and Aven, T. (2009). The role of social cost-benefit analysis in societal decision-making under large uncertainties with application to robbery at a cash depot. *Reliability Engineering and System Safety*, **94** (2009) 1954–1961.

Jones-Lee, M. and Aven, T. (2010). What does the ALARP principle really mean? Revised and resubmitted *Reliability Engineering and System Safety*.

Kahneman, D. and Tversky, A. (1974). Judgment under uncertainty. Heuristics and biases. *Science*, **185**, 1124–1131.

Kaminski Jr., J., Riera, J. D., de Menezes, R. C. R., Miguel, L. F. F. (2008). Model uncertainty in the assessment of transmission line towers subjected to cable rupture. *Engineering Structures*, **30**, 2935–2944.

Kaplan, S. (1991). Risk assessment and risk management – basic concepts and terminology. In *Risk Management: Expanding Horizons in Nuclear Power and Other Industries*. Boston, MA: Hemisphere Publishing Corporation, pp. 11–28.

Kaplan, S. and Garrick, B. J. (1981). On the quantitative definition of risk. *Risk Analysis*, **1**, 11–27.

Keeney, R. L. and McDaniels, T. (2001). A Framework to guide thinking and analysis regarding climate change policies, *Risk Analysis*, **6** (12), 989–1000.

Kettunen, P. (1998). Globalisation and the Criteria of "Us" – A Historical Perspective on the Discussion of the Nordic Model and New Challenges, in Global Redefining of Working Life, Nordic Council of Ministers, Copenhagen.

Knight, F. H. (1921). *Risk, Uncertainty and Profit*. Washington DC: Beard Books. Reprinted 2002.

Kröger, W. (2005). Risk analyses and protection strategies for operation of nuclear power plants. In Landolt-Börnstein New Series Vol. VIII/3B: *Advanced Materials and Technologies/Energy*. Berlin: Springer.

Kröger, W. (2006). Reflections on current and future nuclear safety, *ATW-International Journal for Nuclear Power*, **51**, July, 331–337.

Kujawski, E. and Miller, G. A. (2007). Quantitative risk-based analysis for military counterterrorism systems. *Systems Engineering*, **10**, 273–289.

Kumamoto, H. (2007). *Satisfying Safety Goals by Probabilistic Risk Assessment*. London: Springer.

Kørte, J., Aven, T. and Rosness, R., 2002. On the use of risk analysis in different decision settings. In Proceedings from the ESREL 2002 Conference, 19–20, March, Lyon, France, vol. I, pp. 175–180.

Lauridsen, K., Christou, M., Amendola, A. *et al.* (2001). Assessing the uncertainties in the process of risk analysis of chemical establishments. In *Safety and Reliability. Towards a Safer World*, eds E. Zio, M. Demichela and N. Piccinini, Proceedings. Vol. I. ESREL 2001, Torino (IT), 16–20 Sep 2001. pp. 592–606.

Leveson, N. (2004) A new accident model for engineering safer systems. *Safety Science*, **42**, 237–270.

Leveson, N. (2007) *Modeling and Analyzing Risk in Complex Socio-Technical Systems*. NeTWork workshop, Berlin 27–29 Sept. 2007.

Levy, H. and Sarnat, M. (1994). *Capital Investment and Financial Decisions*. 5th edn. New Jersey: Prentice Hall.

Lindley, D. (1985). *Making Decisions*. New York: Wiley.

Lindley, D. V. (2000). The philosophy of statistics. *The Statistician*, **49**, 293–337. With discussions.

Lindley, D. V. (2006). *Understanding Uncertainty*. New Jersey: Wiley.

Lindley, D. V., Tversky, A. and Brown, R. V. (1979). On the reconciliation of probability assessments (with discussion). *Journal of the Royal Statistical Society A*, **142**, 146–180.

Lirer, L., Petrosino, P. and Alberico, I. (2001) Hazard assessment at volcanic fields: the Campi Flegrei case history. *Journal of Volcanology and Geothermal Research*, **112**, 53–73.

Lowrance, W. (1976). *Of Acceptable Risk – Science and the Determination of Safety*. Los Altos, CA: William Kaufmann Inc.

Luxhøj, J. T., Choopavang, A. and Arendt, D. N. (2001). Risk Assessment of Organizational Factors in Aviation Systems. *Air Traffic Control Quarterly*, **9** (3), 135–174.

Lyse (2007). Capra, G., Cleaver, P., Chester, A. and Phillips, A. QRA of the proposed Lyse Gass LNG base load export terminal, *Advantica, R100-PB-S-SR0001*, 11.04.2007.

Lyse (2008). Quantitative risk analysis (QRA) Lyse LNG base load plant Train 1, *Linde,* R100-LE-S-RS0003, 25.08.2007.

Löfstedt, R. E. (2003). The precautionary principle: risk, regulation and politics. *Trans IchemE*, **81**, 36–43.

Mandel, D. (2007). Toward a concept of risk for effective military decision making. Defence R&D Canada – Toronto. Technical Report. DRDC Toronto TR 2007–124.

Michaels, D. (2008). *Doubt is their Product*. New York: Oxford University Press.

Mohaghegh, Z., Kazemi, R. and Mosleh, A. (2009). Incorporating organizational factors into Probabilistic Risk Assessment (PRA) of complex socio-technical systems: A hybrid technique formalization. *Reliability Engineering and System Safety*, **94**, 1000–1018.

Morgan, M. G. and Henrion, M. (1990). *Uncertainty: A Guide to Dealing with Uncertainty in Quantitative Risk and Policy Analysis*, Cambridge: Cambridge University Press.

Mosleh, A. and Bier, V. M. (1996). Uncertainty about probability: a reconciliation with the subjectivist viewpoint. *Systems, Man and Cybernetics. Part A: Systems and Humans*, **26**, 303–311.

Möller, N., Hansson, S. O. and Person, M. (2006). Safety is more than the antonym of risk. *Journal of Applied Philosophy*, **23**, 419–432.

Nilsen, T. and Aven, T. (2003). Models and model uncertainty in the context of risk analysis. *Reliability Engineering and System Safety*, **79**, 309–317.

North, W. (2010) Probability theory and consistent reasoning. *Risk Analysis*, **30**, 377–380.

NRC (National Research Council) (1983). *Risk Assessment in the Federal Government: Managing the Process,* National Academy of Sciences, Washington, DC: National Academy Press.

NRC (1996). *Understanding Risk: Informing Decisions in a Democratic Society.* National Research Council, Washington, DC: National Academy Press.

NRC (2009). *Science and Decisions: Advancing Risk Assessment.* Washington, DC: National Academy Press.

NRC (2010). Defense-in-depth. US Nuclear Regulatory Commission. www.nrc.gov/reading-rm/basic-ref/glossary/defense-in-depth.html. Accessed 10 February 2010.

O'Brien, M. (2000). *Making Better Environmental Decisions.* Cambridge, MA: The MIT Press.

Östergaard, C., Dogliani, M., Guedes Soares, C., Parmentier, G. and Pedersen, P. T. (1996). Measures of model uncertainty in the assessment of primary stresses in ship structures. *Marine Structures*, **9**, 427–447.

Paté-Cornell, M. E. (1996). Uncertainties in risk analysis: Six levels of treatment, *Reliability Engineering and System Safety*, **54**(2–3), 95–111.

Paté-Cornell, E. and Dillon, R. (2001). Probabilistic risk analysis for the NASA space shuttle: a brief history and current work. *Reliability Engineering and System Safety*, **74**, 345–352.

PSA (2001). *Risk Management Regulations*. Petroleum Safety Authority Norway.

PSA (2002). *The Facilities Regulations. Regulations Relating to Design and Outfitting of Facilities etc. in the Petroleum Activities*, 2002. Petroleum Safety Authority Norway. www.ptil.no/regelverk/category21.html. Accessed 10 May 2010.

PSA (2007). *Guidelines to Regulations Relating to Material and Information in the Petroleum Activities (The Information Duty Regulations)*, §13. Petroleum Safety Authority Norway.

PSA (2009). *Trend in Risk level in the Petroleum Activity*. Summary report 2008. 23.4.2009. Petroleum Safety Authority Norway.

Purple book (2008). Guidelines Risk Calculations (Purple Book) BEVI Module C, Version 3.0 Date 1 January 2008: Modelling specific BEVI categories (BEVI is the abbreviation of the decree implementing the SEVESO directive).

Rasmussen, J. (1997) Risk management in a dynamic society: a modelling problem. *Safety Science*, **27** (2/3), 183–213.

Rechard, R. P. (1999). Historical relationship between performance assessment for radioactive waste disposal and other types of risk assessment. *Risk Analysis*, **19** (5):763–807.

Rechard, R. P. (2000). Historical background on performance assessment for the waste isolation pilot plant, *Reliability Engineering and System Safety*, **69** (3), 5–46.

Reid, S. G. (1992). Acceptable risk. In *Engineering Safety*, ed. D. I. Blockley. New York: McGraw-Hill, pp. 138–166.

Renn, O. (1992). Concepts of risk: A classification. In *Social Theories of Risk*, eds. S. Krimsky and D. Golding. Westport: Praeger, pp. 53–79.

Renn, O. (1998). Three decades of risk research: accomplishments and new challenges. *Journal of Risk Research*, **1** (1), 49–71.

Renn, O. (2008). *Risk Governance*. London: Earthscan.

Research Council of Norway: RCN (2000). Quality in Norwegian Research – An overview of Terms, Methods and Means (In Norwegian only). Oslo.

Rosa, E. A. (1998). Metatheoretical foundations for post-normal risk. *Journal of Risk Research*, **1**, 15–44.

Rosa, E. A. (2003). The logical structure of the social amplification of risk framework (SARF); metatheoretical foundations and policy implications. In *The Social Amplification of Risk*, eds. N. Pidgeon, R. E. Kasperson and P. Slovic. Cambridge: Cambridge University Press, pp. 47–79.

Rosness, R (2009) A contingency model of decision-making involving risk of accidental loss. *Safety Science*, **47**, 807–812.

Ross, S. M. (1993). *Probability Models*, 5th edn. San Diego, CA: Academic Press.

Røed, W., Mosleh, A., Vinnem, J. E. and Aven, T. (2009). On the use of hybrid causal logic method in offshore risk analysis. *Reliability Engineering and System Safety*, **94**, 455–455.

Sahlin, N -E. (1993). On higher order beliefs. In *Philosophy of Probability*, ed. J -P. Dubucs. Dordrecht: Kluwer Academic Publishers.

Sandin, P. (1999). Dimensions of the precautionary principle. *Human and Ecological Risk Assessment*, **5**, 889–907.

Sandin, P., Peterson, M., Hansson, S. O., Rudén, C. and Juthe, A. (2002). Five charges against the precautionary principle. *Journal of Risk Research*, **5**, 287–299.

Saltelli A., Ratto M., Andres T. *et al.* (2008). *Global Sensitivity Analysis: The Primer*. New York: Wiley.

Shafer, G. (1976). *A Mathematical Theory of Evidence*. Princeton: Princeton University Press.

Singpurwalla, N. D. (1988). Foundational Issues in Reliability and Risk Analysis. *SIAM Review*, **30**, 264–282.

Singpurwalla, N. (2006). *Reliability and Risk. A Bayesian Perspective*. New York: Wiley.

Sinn, H. -W. (1980) A rehabilitation of the principle of insufficient reason. *Quarterly Journal of Economics*, **94** (3), 493–506.

Steen, R. and Aven, T. (2010). A risk perspective suitable for resilience engineering. *Safety Science*, **49**, 292–297.

Stirling, A. (1998). Risk at a turning point? *Journal of Risk Research*, **1**, 97–109.

Stirling, A. (2007). Science, precaution and risk assessment: towards more measured and constructive policy debate. *European Molecular Biology Organisation Reports*, **8**, 309–315.

Stirling, A. and Gee, D. (2002). Science, precaution and practice. *Public Health Reports*, **117**(6), 521–533.

Stirling, A., Renn, O. and van Zwanenberg, P. (2006). A framework for the precautionary governance of food safety: integrating science and participation in the social appraisal of risk. In *Implementing the Precautionary Principle*, eds. E. Fisher, J. Jones and R. von Schomberg. Cheltenham: Edward Elgar Publishing, pp. 284–315.

Taleb, N. N. (2007). *The Black Swan: The Impact of the Highly Improbable*. London: Penguin.

Tickner, J. and Kriebel, D. (2006). The role of science and precaution in environmental and public health policy. In *Implementing the Precautionary Principle*, eds. E. Fisher, J. Jones, and R. von Schomberg. Northampton, MA: Edward Elgar Publishing.

US Congress. (2004). Homeland Security: The Balance Between Crisis and Consequence Management Through Training and Assistance. *Hearing before the Subcommittee on Crime, Terrorism, and Homeland Security of the Committee on the Judiciary, House of Representatives, 108th Congress, November* 20, 2003.

US National Research Council (1996). *Understanding Risk*, eds. P. C. Stern and V. Fineberg. Washington D.C.: National Academy Press.

Van Eijndhoven, J. C. M. and Van Ravenzwaaij, A. (2006). Optimizing risk analysis relating to external safety in the Netherlands. *Risk Analysis*, **9**, 495–504.

Vatn, J. (2007). Societal Security – A case study related to a cash depot. In *Risk, Reliability and Societal Safety*, eds. T. Aven and J. E. Vinnem, Proceedings of the European Safety and Reliability Conference 2007 (ESREL 2007), Stavanger, Norway, 25–27 June 2007. London: Taylor & Francis Group, pp. 2599–2607.

Vatn, J. (2010). Issues related to localization of an LNG plant. In *Reliability, Risk and Safety*, eds. R. Bris, C. Guedes Soares and S. Martorell. London: Taylor & Francis Group, vol. **II**, pp. 917–921.

Vatn J, Vatn, G. A, and Drottz-Sjøber, B-M. (2008). Societal security – a case study related to an LNG facility, Social Security Conference, Norwegian Research Foundation.

Vercelli, A. (1995). From soft uncertainty to hard environmental uncertainty, *Economie appliqué'e*, **48**(2), 251–269.

Verma, M. and Verter, V. (2007). Railroad transportation of dangerous goods: Population exposure to airborne toxins. *Computers and Operations Research*, **34**, 1287–1303.

Vinnem, J. E. (2010). Risk analysis and risk acceptance criteria in the planning processes of hazardous facilities – a case of an LNG plant in an urban area. *Reliability Engineering and System Safety.* **95** (6), 662–670.

Vinnem, J. E., Aven, T., Husebø, T., Seljelid, J. and Tveit, O. (2006). Major hazard risk indicators for monitoring of trends in the Norwegian offshore petroleum sector. *Reliability Engineering and System Safety*, **91**, 778–791.

Vose, D. (2008). *Risk Analysis: A Quantitative Guide.* 3rd edn, Chichester: Wiley.

Walley, P. (1991). *Statistical Reasoning with Imprecise Probabilities.* New York: Chapman and Hall.

Weick, K. and Sutcliffe, K. M. (2001). *Managing the Unexpected.* San Francisco: Jossey Bass.

Weinberg, A. M. (1981). Reflections on Risk Assessment. *Risk Analysis,* **1**, 5–7.

Weinberg, A. M. (1972). Science and Trans-science, *Minerun,* **10**, 209–222.

Wiener, J. B. and Rogers, M. D. (2002). Comparing precaution in the United States and Europe. *Journal of Risk Research,* **5**, 317–349.

Willis, H. H. (2007). Guiding resource allocations based on terrorism risk. *Risk Analysis,* **27**(3), 597–606.

Winkler, R. L. (1996). Uncertainty in probabilistic risk assessment. *Reliability Engineering and System Safety,* **85**, 127–132.

Wolfs, F. (2009). Introduction to the Scientific Method – An explanation on what the scientific method is and does. http://teacher.nsrl.rochester.edu/phy_labs/ AppendixE/AppendixE.html. Accessed 3 March 2010.

Zio, E. (2009). Reliability engineering: Old problems and new challenges. *Reliability Engineering and System Safety,* **94**, 125–141.

Zio, E. and Apostolakis, G. E. (1996). Two methods for the structured assessment of model uncertainty by experts in performance assessments of radioactive waste repositories. *Reliability Engineering and System Safety,* **54**, 225–241.

Index